Praise for *My River Chronicles: Rediscovering the Work that Built America*

"[DuLong] delivers an engaging narrative of maritime history and her own hands-on perceptions"

—*The New York Times*,

"An unexpected portrayal of America in the decline of industry, delivered from the unique vantage point of the Hudson River.... an eye-opening picture of what America has been ... and what it is becoming ... Powerful."

—*Kirkus*, starred review

"She details her often exhilarating experiences in her very fine and gutsy book. Ms. DuLong is a confident and sensual writer, as perceptive about small matters on a boat as was Anthony Bourdain, in *Kitchen Confidential*, about everyday events in a professional kitchen."

—*The New York Times*, ArtsBeat blog

"Smart, captivating prose ... [Readers] will love this unusual mix of history, adventure, feminism, and blue-collar know-how. Highly recommended."

—*Library Journal*

"Jessica DuLong's elegantly written *My River Chronicles* ... carries forward the craft of literary non-fiction with grace and energy."

—Gay Talese, *A Writer's Life*

"In rich and captivating prose, Jessica DuLong kindly invites the rest of us on the journey of her lifetime: from a dot-com job to the fabled waters of the Hudson River, where she became a fireboat engineer. This is an unusual and fascinating book."

—Jon Meacham, *American Lion*

"When Jessica DuLong describes her work in the engine room of the *John J. Harvey*, you can practically feel the throb of the boat's mighty diesels. This is someone who has paid some dues, and it shows in the details. Her view through a narrow portal at the water line opens into a bigger picture of the Hudson River, the economy of New York, and the dignity of work—the kind of work that is genuinely useful. *My River Chronicles* is an account of what made this country thrive, and might yet again: men and women who aren't content to stand around with their hands in their pockets. The book reeks of penetrating oil, which may be just what is needed to get our economy, and our culture, moving again."
—Matthew B. Crawford, *Shop Class as Soulcraft*

"Jessica DuLong is a lucky woman. She stumbled into an obscure world—the overheated engine room of an old fireboat—and discovered that she belonged there. Readers are lucky, too, because she has managed to translate her love affair with the water into a finely written and fascinating story about a lost American way of life."
—Stefan Fatsis, *Word Freak*

"In a world where we are growing increasingly disconnected from anything real, what a delight to enter the engine room with Jessica DuLong, a real person doing a real thing in a real place. This is the kind of river trip that memoirs were developed for in the first place."
—Douglas Rushkoff, *Present Shock*

"As both a writer and an engineer, [DuLong is] relentlessly, gratifyingly curious, and her fine, richly detailed prose holds an appeal regardless of your level of interest in heritage histories and engine mechanics.... DuLong's passion for her craft is contagious, making *My River Chronicles* one of the most moving, unusual books I've read in a long time."
—*Bookslut*

DUST TO DELIVERANCE

Untold Stories from the Maritime Evacuation on September 11

JESSICA DuLONG

New York Chicago San Francisco Athens London Madrid
Mexico City Milan New Delhi Singapore Sydney Toronto

To my parents, Peter and Gretchen,
*whose lessons about language and
goodness charted my course*

To Ben, Zillin, and Jude,
for buoying me with love and light

Copyright © 2017 by Jessica DuLong. All rights reserved. Printed in the United States of America. Except as permitted under the United States Copyright Act of 1976, no part of this publication may be reproduced or distributed in any form or by any means, or stored in a database or retrieval system, without the prior written permission of the publisher.

1 2 3 4 5 6 7 8 9 0   QFR   21 20 19 18 17

ISBN   978-0-07-180498-1
MHID   0-07-180498-6

eISBN   978-0-07-180499-8
eMHID   0-07-180499-4

McGraw-Hill Education books are available at special quantity discounts to use as premiums and sale promotions, or for use in copyright training programs. To contact a representative, please visit the Contact Us pages at www.mhprofessional.com

Dust to Deliverance: Untold Stories from the Maritime Evacuation on September 11 has been funded in part by Furthermore grants in publishing, a program of the J.M. Kaplan Fund.

Photos used with permission of the New York City Police Department ©2001.

Contents

September 11, 2011 Timeline   vi

PART ONE   **The Situation**
CHAPTER 1:   "It was a jet. It was a jet. It was a jet!"   3
CHAPTER 2:   "Shut it down! Shut it down!"   22
CHAPTER 3:   "NEW YORK CITY CLOSED TO ALL TRAFFIC"   41

PART TWO   **The Evacuation**
CHAPTER 4:   "I was gonna swim to Jersey."   54
CHAPTER 5:   "It was like breathing dirt."   71
CHAPTER 6:   "We're in the water!"   86
CHAPTER 7:   "Gray ghosts"   97
CHAPTER 8:   "A sea of boats"   114
CHAPTER 9:   "I need a boat."   134

PART THREE   **The Aftermath**
CHAPTER 10:   "We have to tell us what to do."   160
CHAPTER 11:   "Sell first, repent later."   178
CHAPTER 12:   "Thanks for your help!"   191
CHAPTER 13:   "They'd do it again tomorrow."   199
CHAPTER 14:   September 11, 2016   208

Afterword   218
Acknowledgments   220
Vessel Participants   222
Notes   225
Index   231

September 11, 2001 Timeline

A.M.

8:46:30	American Airlines Flight 11 crashes into 1 World Trade Center (the North Tower)
8:47	MTA subway operator alerts MTA Subway Control Center of an explosion in the WTC and begins emergency procedures
8:52	PATH trains begin emergency procedures and proceed to evacuate WTC station and return Manhattan-bound trains to New Jersey
9:02:59	United Airlines Flight 175 crashes into 2 World Trade Center (the South Tower)
9:10	U.S. Coast Guard closes Port of New York and New Jersey Port Authority of NY and NJ closes all its bridges and tunnels
9:17	FAA shuts down all NYC airports
9:37:45	American Airlines Flight 77 hits the west wall of the Pentagon
9:58:59	South Tower collapses
10:03:11	United Airlines Flight 93 crashes in Pennsylvania
10:20	NYC Transit suspends all subway service
10:28:22	North Tower collapses
10:30	NJ Transit stops rail service into Manhattan's Penn Station
10:45	U.S. Coast Guard calls for "all available boats" to assist evacuation of Lower Manhattan; PATH operations suspended
11:02	Mayor Giuliani calls for evacuation south of Canal Street

P.M.

5:20	7 World Trade Center, headquarters of NYC Office of Emergency Management, collapses

Sources:

U.S. Department of Transportation's John A. Volpe National Transportation Systems Center, "Effects of Catastrophic Events on Transportation System Management and Operations: New York City—September 11," 2/6/17, https://ntl.bts.gov/lib/jpodocs/repts_te/14129.htm.

U.S. Department of Commerce, Technology Administration, National Institute of Standards and Technology, "Final report on the collapse of the World Trade Center towers," 4/26/17, http://ws680.nist.gov/publication/get_pdf.cfm?pub_id=909017.

P.J. Capelotti, *Rogue Wave: The U.S. Coast Guard on and after 9/11*, Washington, DC, U.S. Coast Guard Historians Office, 4/27/17, https://www.uscg.mil/history/91101/pubs/911_Rogue_Wave.pdf.

PART ONE

THE SITUATION

There is a crack in everything.
That's how the light gets in.
—Leonard Cohen

CHAPTER 1

"It was a jet. It was a jet. It was a jet!"

AS THE SUN TRACKS ACROSS THE SKY ON THIS OVERCAST, 78-degree morning, the clouds part ways leaving behind a mazarine blue. There is no dust. No smoke. The heaviness in today's air is only the humidity of late summer. A forest of sailboat masts bobs in the rectangular notch of Manhattan's North Cove. The propeller wash from New York Waterway and Liberty Landing ferries dropping off and picking up passengers at the new World Financial Center terminal, 150 paces or so to the north of the small harbor, pushes little waves through the 75-foot gap in the breakwater. *Mis Moondance*, a 66-foot charter yacht, motors in and maneuvers into a slip among the wooden floating docks. A blue and white police boat holds station just outside the cove's entrance, blue light flashing above the pilothouse.

To a casual observer, unaware of the date, it might be hard to say if this quiet is just the regular hush of Sunday or something more solemn. Certainly the pedestrian plaza is far less populated on Sundays than it would be on a weekday morning—a Tuesday morning, say. Surely all the street closures and police barricades thwarting access have kept some people away, while reminding those that might have momentarily forgotten that this is no ordinary day.

Several blocks inland, beneath the trees in the National September 11 Memorial plaza, the fifteenth anniversary commemoration has begun. About 8,000 people have assembled for this year's annual ritual. Families of those lost will read, 30 at a time, the names of the 2,977 people who died from injuries or exposures sustained 15 years ago today, plus the six killed in the bombing of the World Trade Center on February 26, 1993.

At 8:46 A.M., bells ring in the plaza and across New York City to announce the first of six moments of silence. This first marks the moment when American Airlines Flight 11 careened through the northern facade of the World Trade Center's North Tower between the ninety-third and ninety-ninth floors. By the water's edge, the chuff-chuff-chuff of a helicopter hovering over the Hudson never lets up. Silence on the waterfront is merely theoretical.

Sunday joggers, earbuds in, digital music players strapped around biceps, continue on their morning runs. Bicyclists keep biking, tourists snap photographs, parents herd young children. But two New York Waterway ferries pause, foregoing their usual over and back, over and back, to linger in reverence. Above them glints the new 1 World Trade Center, the base of its spire reflected in an adjacent skyscraper, also new. Between them stands a third tower, still under construction, the outstretched arm of a crane loitering above its uppermost reaches—a skeleton waiting for workers to finish grafting on its reflective skin.

When the bell chimes again at 9:03 A.M., the moment that United Airlines Flight 175 crashed into the South Tower's southern facade between the seventy-seventh and eighty-fifth floors, a lone gentleman with close-cropped gray hair sits silently before the cross inside St. Joseph's Chapel where a special anniversary mass is scheduled to commence at 10 A.M., one minute after the South Tower fell, and 28 minutes before the North Tower followed it to the ground.

The day after the attacks, this chapel was converted into a makeshift, volunteer-run supply house for distributing donated goods. Rescue workers turned its plate glass windows into a message board of sorts, tracing pleas, prayers, and pronouncements into the gray dust: "Revenge is sweet." "Goodness will prevail." "It doesn't matter how you died, it only matters where you go." "You woke a sleeping giant." Among the scrawls was the word "*Invictus.*" Latin for *unconquerable*, it's the title of an 1875 poem by William Ernest Henley that begins:

"Out of the night that covers me,
  Black as the Pit from pole to pole,
  I thank whatever gods may be
  For my unconquerable soul."

Other messages, more practical than poetic, included: "Go to Stuyvesant High School to sleep" and "Lt. John Crisci call home."

I first transcribed these missives into a small reporter's notebook while standing in the dust of September 12, 2001. Although I wasn't technically reporting at the time, a writer's lifelong habits run deep. I scribbled down the words in an attempt to collect the details that I hoped might somehow help me make sense of the unfathomable ruination at hand. At 28 years old, I was still a newcomer to New York City, having moved here in January of 2000. By the following September I was just six months into the hands-on apprenticeship that had launched my new career as a marine engineer. I was a novice in every sense of the word.

Now, a decade and a half later, I've risen from assistant engineer to chief, a "hawsepiper" who's come up through the ranks (climbing, metaphorically, up the anchor chain through an opening in the bow called the hawsepipe) by learning on the job rather than in school. New York harbor's maritime community is my community.

After 15 years in the industry, my view of everything has changed. Now, on this overcast Tuesday morning, when I notice the dull red paint coating one section of the curved steel railing along the water's edge, I recognize it as primer, evidence of a painting project in process. This is the railing that I climbed over, on September 12, as I bolted from threats of a fourth building collapse, scrambling to board the boat that had so recently become my workplace: retired 1931 New York City fireboat *John J. Harvey*. No longer an active-duty vessel with the Fire Department of the City of New York (FDNY), the boat had been operating as a preservation project and living museum when it was called back into service to help fight New York City's most devastating fires.

A fireboat is essentially a huge pump, several pumps, in fact, which is exactly what was needed that day. And so fireboat *Harvey*'s all-volunteer, all-civilian crew (save for our captain, a retired FDNY pilot) worked alongside active-duty fireboats to pump Hudson River water to land-based battalions. Fire mains lay broken. Hydrants were buried beneath debris. For days following the twin towers' collapse, fireboats provided the only firefighting water available on site. When firefighters bent over their hoses to rinse the dust from their faces, they sputtered and spit in surprise at the taste of salt from the Hudson.

Supporting pumping operations aboard fireboat *John J. Harvey* was the work that had brought me to Ground Zero. I'd spent the eleventh like so many others: glued to the television, then wandering around my Brooklyn neighborhood trying unsuccessfully to donate blood. My identity as a mariner was not yet ingrained. As I'd watched the staticky news coverage on the only channel that would come in on my television set, it hadn't occurred to me that the antique, decommissioned fireboat where I was spending more and more of my days as a budding engineer could offer the opportunity to help that I sought so desperately. And so I missed the boat lift.

As thick, gray smoke began spilling through the airplane-shaped hole in the World Trade Center's North Tower, civilians caught in an act of war—some burned and bleeding, some covered with soot—fled to the water's edge, running until they ran out of land. Never was it clearer that Manhattan is an island. Within minutes, mariners had raced to meet them, white wakes zigzagging across the harbor. Long before the Coast Guard's call for "all available boats" crackled out over marine radios, scores of ferries, tugs, dinner boats, sailing yachts, and other vessels had begun converging along Manhattan's shores. Hundreds of mariners shared their skills and equipment to conduct a massive, unplanned rescue. Within hours, nearly half a million people—men, women, and children—had been delivered from Manhattan by boat.

This became the largest waterborne evacuation in history—more massive even than the famous World War II rescue of troops pinned by Hitler's armies against the coast in Dunkirk, France. In 1940, hundreds of naval vessels and civilian boats rallied to rescue 338,000 British and Allied soldiers over the course of nine days. But on September 11, 2001, boat crews evacuated an estimated 400,000 to 500,000 civilians in less than nine hours. The speed, spontaneity, and success of this effort were unprecedented.

In the years since, countless shattered lives have been remade, fractured families reconstructed, loves lost and found. "The Pile"—16 acres of wreckage left at the World Trade Center site—was eventually excavated and redubbed "the Pit" before being transformed into a memorial with twin reflecting pools that occupy the square footprints of the vanished towers. Americans' pre-9/11 sense of security, along with a misbelief in our immunity to the carnage and cruelty suffered by the rest of the world, was sabotaged and replaced with a gnawing "new normal." This post-9/11 "afterward" was characterized by anxiety and suspicion coupled with an acquiescence to new infringements on privacy and freedom. But what also arose in the aftermath of the deadliest terrorist attacks on U.S. soil was a heightened sense of goodwill, an abundance of comity, and a reflexive compulsion to help. Amid the darkness and chaos, a series of lifesaving, selfless acts transformed the waterfront of New York harbor into a place of hope and wonder.

By the time I arrived at the trade center, tugs and other vessels lining the seawall had shifted gears from ferrying people to running supplies and other critical support operations. In the hazy, horror-filled, dust-choked days that followed, I didn't grasp that history had been made along Manhattan's shores. Indeed, even today few people recognize the significance of the evacuation effort that unfolded on that landmark day. This book addresses that omission. The stories that follow are the culmination of nearly a decade of reporting to discover how and why this remarkable rescue came to pass—what made the boat lift necessary, what made it possible, and why it was successful.

On any given Tuesday in 2001, you could stand at the southern tip of Manhattan Island, gaze out over the water, and watch the busyness of New York harbor unfold before your eyes. You'd doubtless notice a Staten Island Ferryboat, in all its enormous orange glory, bridging the 5.2-mile gap between the two island boroughs. Looking up the west side, you'd see smaller white and yellow fast ferries darting across the three-quarter-mile span of the Hudson that separates Manhattan from New Jersey. Maybe you'd track the movements of a recreational sailor, playing hooky on a weekday, tacking back and forth through the sparkling salt water to drink in the last of the summer sun. Over toward the east side, on the approach to the half-mile expanse of the East River between Manhattan and Brooklyn, you might lay eyes on a black-hulled freighter making its way to tie up at the Brooklyn Navy Yard.

Scanning across the waters straight ahead, your view bracketed by container cranes in Brooklyn's Red Hook Terminal on the left and Port Elizabeth's and Port Newark's on the right, you'd perhaps catch glimpses of the working harbor carrying out its workaday business: a tugboat pushing barges filled with scrap metal or stone; another tug in the anchorage securing empties to await a fair tide; a North River-bound bulk freighter, the booms of its white deck cranes outstretched like dueling swords; a containership, nudged along through the channel by ship-assist tugs. Or maybe you'd even spot a "honey boat" hauling sewage sludge from local wastewater treatment plants, or an Army Corps of Engineers drift collection vessel plucking wreckage from the water to remove hazards to navigation. Together these watercraft, working side by side, under the oversight of the U.S. Coast Guard, perform the critical functions of the Port of New York and New Jersey.

Still, much of the activity of New York harbor would remain unnoticed. A quick tour of the numbers reveals how much activity there was in September of 2001. Back then, New York harbor provided passage to 91,600 commuters and accommodated between 25 and 30 large, international, deep-draft, commercial vessels on an average weekday. Including 30 billion gallons of petroleum and petroleum products, more than $93 billion worth of cargo moved through the port annually, generating a total of $29 billion in economic activity while serving more than 17 million customers in the states of New York and New Jersey. More than 167,000 people made their living directly from all this traffic.

New York harbor was, and is, a busy place—the third largest container port in the United States and a vital connection between New York City and the rest of the world. But other than the passenger ferries, whose crews interface directly with their customers, much of the hard work of the harbor's working watercraft happens—now, as it has since the latter half of the twentieth century—largely out of view. Manhattan is an island, and the realities of island real estate are what ushered the port's industries off Manhattan's shores and over to Brooklyn, Staten Island, and New Jersey in the 1960s and '70s.

Although port workers handled nearly every item in New Yorkers' home and work lives on its arrival from overseas, most residents hardly gave the harbor a passing thought. By late 2001, the last vestiges of the borough's working waterfront had been rapidly uprooted and replaced with sparkling esplanades festooned with iron railings and polished stone. Maritime infrastructure (cleats, bollards, fendering, and other features necessary for a safe tie-up) had been replaced with ornamental fencing. An island that had once berthed legions of vessels now had a waterfront that was mostly geared toward recreation and people's enjoyment of a passive view.

On September 11, 2001, as the cascade of catastrophe unfolded, people found their fates altered by the absence of that infrastructure and discovered themselves dependent upon the creative problem solving of New York harbor's maritime community—waterfront workers who'd been thrust beyond their usual occupations and into the role of first responders.

"It was a jet. It was a jet. It was a jet!" The time was 8:46 A.M.

Patrick Harris had been sipping coffee at the open helm of his 63-foot wood sailing charter yacht *Ventura*, chatting with the boat's maintenance man, when he heard the roar of engines up close. He looked up in time to watch the tail of a jet airplane penetrate the north face of 1 World Trade Center, less than 1,000 feet away from where his boat was tied to a floating dock in North Cove, the small harbor notched out of Manhattan's western shore where the World Financial Center meets the Hudson.

"In my mind's eye I can still see a frozen Kodachrome of the tail end of the aircraft—the rudder mechanism and the back fifth of the plane—disappearing into the building," he explained. "As soon as it slammed in, there was total silence."

Two seconds later, five stories of windows lit up in bright orange "like a pinball machine." Harris heard a whoosh like a barbecue grill igniting and saw a fireball blast through the north face of the building. A "big, black billowing cloud with orange flashes in it" burst into the clear blue sky.

The captain sat stunned for a moment before reaching for his marine VHF radio, the standard radio equipment installed aboard vessels large and small that operates over the "very high frequency" maritime mobile band. With one hand on the wheel, he pulled the microphone off its clip and depressed the button with his thumb. "United States Coast Guard, Group New York. Sailing vessel *Ventura* on one-six."

"U.S. Coast Guard Activities New York to the *Ventura*."

Here, Harris froze. The maintenance man had repeated the word three times in the instant after impact, confirming what Harris had witnessed. But now the captain couldn't bring himself to say the word *jet*. "There's been a tremendous explosion at the World Trade Center," he said. "It looks like five to eight stories are on fire. You're going to need some backup here."

All protocol fell away. "What?!?" said the youthful voice on the other end of the transmission. In that instant, Harris felt the shift as the formality of a vessel captain calling the Coast Guard vaporized.

"It looks like a plane hit," he continued, haltingly, explaining that he planned to head on foot toward the towers with his handheld and would radio back with whatever information he could gather.

Harris, calling just seconds after American Airlines Flight 11 had rammed into the North Tower, was the first to notify the U.S. Coast Guard of the unfolding disaster.

Word traveled quickly up the chain to the deputy commander of the Coast Guard's Activities New York, who, as it happens, was also named Patrick Harris. That morning Rear Admiral Richard E. Bennis was out of town, leaving Deputy Commander Harris as the acting captain of the Port of New York and New Jersey, the Coast Guard's largest operational field command. At the Coast Guard station in Fort Wadsworth on Staten Island, a watch stander interrupted the morning meeting with: "Hey, Captain, I think you oughta take a look at this."

Harris stepped a few feet away into the Vessel Traffic Service (VTS) center, a dark room full of monitors, phones, VHF radios, and radar screens glowing brightly as they offered a detailed overview of more than 40 nautical square miles. Twenty-four hours a day, seven days a week, operators manned this maritime equivalent of an air traffic control center, monitoring the movements of vessels traveling through the Port of New York and New Jersey. To prevent vessel collisions and groundings, expedite ship movements, increase transportation system efficiency, and improve operating capabilities in all weather, they tracked traffic by remote radar and live television cameras mounted strategically throughout the harbor, and communicated with each vessel via phone or VHF. On a typical day the VTS supervised and assisted about 750 large vessels passing through the port.

Suddenly, this was no typical day. All eyes were trained on the live feed from a dozen of the 16 strategically placed cameras that showed 1 World Trade Center, the North Tower, burning. On the television in a nearby break room, CNN reported that a small plane had hit the tower. This news left the deputy commander, a career aviator with 34 Air Medals from his Vietnam service, uneasy. As a pilot, he couldn't believe that a small plane could have done this kind of damage.

So far, in the minutes before 9 A.M., the incident was land-based, not technically of Coast Guard concern, but a 41-foot search and rescue utility boat was dispatched to the scene, just in case. Harris grabbed the phone and dialed Rear Admiral Bennis who, at the time, was in Northern Virginia cruising down I-95 South with his wife. Get to a television, Harris advised. "We don't know what happened, but a plane just hit one of the towers of the World Trade Center."

One hour earlier, Tammy Wiggs had walked across the five-acre, stone-paved plaza at the heart of the World Trade Center complex on her way to the New York Stock Exchange. The newly minted Georgetown University graduate had held her new job as a Merrill Lynch clerk for only a week and a day. Although she wasn't expected at 11 Wall Street until after eight, her determination to make a name for herself straight out of the gate had prompted her to leave her Upper East Side apartment at 6:15 A.M. for the chance to show her face, walk around, shake hands, and chat up the traders at the desk in 4 World Financial Center. She planned to *be* one of those traders one day. While most clerks stopped in for a meet-and-greet one or two mornings per week, 22-year-old Wiggs, a self-described "naturally competitive person," planned to outshine them all with daily visits. Everyone would remember her face.

At about quarter to eight she'd finished making her rounds and headed outside. Relishing the bright morning sun, she crossed the plaza at the foot of the two 110-story towers that had claimed title, at their 1973 ribbon-cutting ceremony, as the world's tallest buildings. The twin towers had since become the nerve center of the bustling financial district, a must-see tourist attraction, and a central focus of the New York skyline.

Along her way, Wiggs passed planter boxes, stone benches, a stage set for summer performances, and the public plaza's cynosure at the center of a spill-over fountain: a 25-foot-high bronze sphere sculpture, created by German artist Fritz Koenig, that symbolized world peace through world trade. This was a fitting centerpiece to the complex that architect Minoru Yamasaki had designed as "a living symbol of man's dedication to world peace" and a "representation of man's belief in humanity."

Now 50,000 people worked in the vast, seven-building World Trade Center complex that housed some 1,200 companies and organizations. Featuring restaurants, a shopping mall, and a belowground transit hub, the site saw an average of 90,000 visitors daily. Just six weeks earlier, the Port Authority of New York and New Jersey had leased the property, now more than 99 percent occupied, to a private developer, Larry Silverstein, who'd agreed to pay the equivalent of $3.2 billion over the next 99 years.

But none of that was on Wiggs's mind as she made her way to Wall Street. Ten minutes later, with a heady mix of exhilaration and fear, she stepped into the New York Stock Exchange coatroom, swapped heels for Steve Madden loafers, and donned her black, mesh-backed floor jacket with the Merrill Lynch bull patch on the sleeve. Then she passed through the turnstile and strode to her station in a room dubbed "the garage."

Up close, the stock exchange floor looked in real life just like it did on television: a cluster of round islands littered with computer keyboards and blinking screens. What struck Wiggs most about the space, which at this hour thrummed with pre-opening-bell anticipation, was its gymnasium vastness. People milled about issuing good mornings, coffee cups in hand. Wiggs felt like the new kid—female at that—walking into a guys' locker room.

She was standing in her spot at one of Merrill Lynch's booths, checking the "breaks" to make sure all of Monday's trades had gone through correctly, when she heard a thump that sounded like a truck hitting a building. It was 8:46 A.M. A murmur rose across the floor as television screens lit up with announcements that a small plane had struck the North Tower. At first the chatter centered around how this new development might fit into the trading day. But then, at 9:03 A.M., a collective gasp erupted.

Across the Hudson River, about five miles to the north, Michael McPhillips was working aboard a converted car ferry docked in Weehawken, New Jersey, that now housed offices for the ferry company New York Waterway. A lifelong mariner, he'd run away to sea at 16 and traveled the world aboard ships before taking a position as Waterway's port captain, charged with supervising vessel deployment, terminal oversight, management of captains and deckhands, vessel maintenance, and Coast Guard compliance. He was busy checking the day's schedule, making sure the ferryboats were running smoothly and on time, when a deckhand called to tell him the World Trade Center was on fire.

He drove a half-mile to the work dock and corporate offices, dialing his superiors en route. After coordinating with company vice president Donald Liloia and rounding up two mechanics to work as deckhands, McPhillips commandeered the ferry *Frank Sinatra*, which had been out of service with one of its four engines down. Confident that he could adjust the throttles to compensate for the downed engine, McPhillips pulled the boat off the dock and shot straight for the World Financial Center. As he stood in the wheelhouse, it looked like the whole top of the North Tower was afire.

New York Waterway, which carried 32,000 passengers on an average weekday, had no plan for what to do if a plane struck a trade center tower. But the company did have a protocol for dealing with service interruptions of the PATH train—the Port Authority Trans-Hudson Corporation's commuter rail connecting New Jersey to Manhattan through tunnels under the Hudson River. Since the railway terminated right beneath the World Trade Center complex, there was little doubt that a fire would provoke a PATH shutdown, leaving New Jersey commuters hunting for other ways to get home. And so Waterway went immediately into stopgap mode to offer alternative transportation.

"Deployment was a piece of cake," McPhillips recalled years later. It was rush hour, so most of the boats were already crewed-up and operating. Instead of letting the boats wind down, as they usually did by late morning, the managers ordered crews to keep running.

The white-tent-covered barge that functioned as the World Financial Center's ferry-loading platform was swarming with people when McPhillips dropped off Liloia. The vice president would manage passenger boarding by land; McPhillips would oversee operations by water. The boats would work overtime until train service was restored. *This plane crash was a terrible accident.*

Just a fire. Everything will settle down by evening rush hour. With this thought, McPhillips set about leading ferryboat captains in transporting boatloads of people off the island.

Minutes later, Waterway's director of operations Peter Johansen also arrived at the ferry terminal. He'd been standing in the wheelhouse of a company catamaran ferry bound from Weehawken to Pier 11 on the East River when he'd caught sight of the jet getting "sucked into" the North Tower. As the boat continued around the southern tip of Manhattan, Johansen could see that the force of the plane's impact had blown out the windows on the opposite side of the tower. Smoke now poured out of the south face of the building as well. "Honestly, I think most people felt that it was a navigation accident. The reason I say that is because we continued around to Pier 11, the Wall Street terminal, and there were about a hundred people on board. Every single one of them got off and went to work that morning. And as they're walking off there are envelopes and letters floating down from the sky." At about 8:55 on a Tuesday morning, no one on that boat could have foreseen the scale of the trouble to come.

Unlike those commuters, Johansen had dropped his day's plans and rerouted. Instead of attending a meeting of the Port of New York and New Jersey's Harbor Safety, Navigation, and Operations (Harbor Ops) Committee, he'd directed a ferry to drop him at the World Financial Center to help manage the increased ferry traffic that would doubtless result from the incident.

When Johansen arrived at the ferry terminal, he and Liloia split duties, with Johansen manning the top of the gangway, permitting only as many passengers as could fit on the next boat to go down onto the barge, and Liloia at the bottom, designating which boats would go where. "Everybody was standing there. Nobody panicked," Johansen recalled. "In the beginning we were taking people to their regular stops. . . . Later on it was just, 'Get 'em across the river.' So it was either Hoboken or across to Jersey City." By midnight New York Waterway would use more than 20 different ferryboats to transport more than 160,000 people.

Less than a mile away from the World Financial Center terminal, FDNY Battalion Chief Joseph Pfeifer had seen the first plane hit and seconds later he called in to FDNY's Manhattan Dispatch with both a first and a second alarm. Racing south in his battalion car, he realized that the 19 trucks those alarms summoned would not be enough, and he called in a third. "We have a number of floors on fire," he explained. "It looked like the plane was aiming towards the building." Reports from other units echoed his sense that this collision was no accident. They provided the first glimpses of the mounting disaster, painting a doomish picture.

8:48:09

Engine 1-0: *It appears an airplane crashed into the World Trade Center....*

Squad 1-8: *... looked like it was intentional. Inform all units coming in from the back it could be a terror attack....*

Engine 1-0: *Roll every available ambulance you've got to this position....*

Engine Fire 5: *Please have ambulances respond to West Street, we have several injured people on West Street here....*

Pulling up in front of the North Tower at 8:50 A.M., Chief Pfeifer was the highest-ranking fire commander on scene. Looking up the west side of the building, he saw light smoke but no fire. He swapped his white chief's hat for his fire helmet and yanked on his protective bunker gear—overalls, coat, boots—before stepping into the lobby to confront the biggest conflagration of his career. Shattered glass from blown out windows crunched under his thick rubber boots, and the first of one thousand firefighters summoned began streaming into the tower to report to the largest rescue operation in New York City history.

Although they couldn't see what the fire and impact zone looked like from the outside, the firefighters in the lobby of the

North Tower knew they were in deep trouble. The size of the gash and the intensity of the smoke and flames were beyond the fire-extinguishing capability of the forces they had on hand.

Minutes after Chief Pfeifer reached the North Tower, the division chief of Lower Manhattan Peter Hayden arrived on scene and took over. As Hayden later explained: "We realized that, because of the impact of the plane, there was some structural damage to the building, and most likely that the fire suppression systems within the building were probably damaged and possibly inoperable."

Here, simple physics reigned supreme. Each single trade center floor stretched 40,000 square feet—nearly an acre. Even with multiple hose lines, each capable of dousing 2,500 square feet, the FDNY could not battle a fire that had already engulfed at least five floors. Very early on the chiefs determined that this was strictly a rescue mission. "We were going to vacate the building, get everybody out, and then we were going to get out. . . . We had a very strong sense that we would lose firefighters . . . but we had estimates of 25,000 to 50,000 civilians, and we had to try to rescue them."

This monumental decision meant that all the companies now crowding the lobby to muster at the scene of what would become the deadliest fire in U.S. history would not, in fact, attempt to fight the blazes. Instead they would focus on evacuation. Thwarting their efforts, however, was the fact that almost none of the building's 99 elevators still functioned. To reach the upper floors, firefighters, laden with between 56½ and 94½ pounds of gear, would have to take the stairs, as would evacuees coming down.

Eight years earlier, elevator failures had plagued rescue workers, as well. At 18 minutes after noon on February 26, 1993, a huge bomb had exploded beneath the World Trade Center, killing six people (one of them pregnant) and injuring more than 1,000. Terrorists had planted a 1,500-pound bomb on a timer in a rented Ryder truck and parked it on Level B-2 of the underground garage.

The subsequent blast had opened a hole seven stories deep, cutting off electrical power and communications in significant areas of the twin towers above. The public-address system and even the emergency lighting system had failed. People (including a teacher and a group of schoolchildren) had ended up trapped for hours in elevators while evacuees using cigarette lighters made their way down dark and smoky stairways. Evacuation of both towers had taken more than four hours. "Without elevators," Donald Burns, a chief on the scene of the 1993 bombing, had written in an after-action commentary, "sending companies to upper floors in large high-rise buildings is measured in hours, not minutes." Now, the impact of the plane left rescue workers struggling once again with major failures in building infrastructure.

Damage caused by the jetliner collision had left some elevator cars, filled with trapped riders, stalled between floors. Other cars had reached the lobby level but failed to open. Screaming passengers banged on the sealed doors just feet away from firefighters gathering in the lobby. But amidst the sirens, the shouting, and the clamor of competing radios, their screams went unheard.

Also mobilizing were officers from the City of New York Police Department (NYPD) and the Port Authority Police Department (PAPD), charged with protecting the Port Authority's customers, commuters, and employees. Ten minutes after the first plane hit, at 8:56 A.M., NYPD Chief of Department Joseph Esposito had radioed central dispatch calling for a "Level 4" mobilization, the department's highest state of alert, equivalent to a "war footing." Level 4 marshaled nearly 1,000 officers to the scene— 22 lieutenants, 100 sergeants, and 800 police officers from all over the city. Already an elite team of about 40 rescue specialists from the NYPD's Emergency Service Unit (ESU) had arrived and set up a command post at Church and Vesey Streets on the northeastern corner of the trade center complex, right across the street from a Borders bookstore where a children's story hour was set to begin.

Within about 15 minutes, New York City and the Port Authority of New York and New Jersey launched the largest rescue operation in the city's history, deploying first responders, beginning an evacuation, and making the critical decision to focus efforts in the North Tower on rescue and not firefighting.

———⊖⊗⊗⊖———

Meanwhile, the U.S. military, having received information about the hijackings too late to change the course of events, played catch-up as the attacks unfolded. Although the Federal Aviation Administration (FAA) Boston Center flight controllers had learned by 8:20 A.M. that Flight 11 had probably been hijacked, the U.S. military wasn't informed until 17 minutes later when, at 8:37 A.M., an air traffic controller at the FAA's Boston Center called the Northeast Air Defense Sector (NEADS), a division of the North American Aerospace Defense Command, a binational U.S. and Canadian command charged with defending North American airspace.

FAA: *We have a problem here. We have a hijacked aircraft headed towards New York, and we need you guys to, we need someone to scramble some F-16s or something up there, help us out.*

NEADS: *Is this real-world or exercise?*

FAA: *No, this is not an exercise, not a test.*

At 8:46, NEADS ordered to battle stations two F-15 Eagle fighters from Otis Air National Guard Base in Falmouth, Massachusetts, 153 miles away from New York City. The hijackers aboard American Airlines Flight 11 had turned off the plane's transponder, and NEADS personnel were still searching their radar scopes for the airliner at 8:50 when word reached them that a plane had struck the North Tower four minutes prior.
Before the plane hit, Flight 175's pilots had reported a "suspicious transmission" during which a Flight 11 hijacker keyed the wrong microphone and widely broadcast his announcement to passengers to remain seated. Shortly thereafter, between 8:42 and 8:46 A.M., their own plane was seized. The military remained unaware of the second hijacking. Although fighter jets got

airborne at 8:53 A.M., they had no known target and were sent to military-controlled airspace off the coast of Long Island.

On the ground in Lower Manhattan, the rescue operation was taking shape. While first responders raced into the complex, civilians raced out. By 8:46 on the morning of September 11, an estimated 16,400 to 18,800 people were present in the World Trade Center complex. And, of course, the surrounding area was filled with homes and businesses as well. Lower Manhattan, the area south of Chambers Street and the Brooklyn Bridge, was home to 22,700 residents. It was also the fourth largest business district in the nation, with a private sector working population of 270,200 in addition to public sector employees. With the inclusion of tourists, students, shoppers, and other visitors, the area's daytime population swelled into the hundreds of thousands.

New York City is the most densely populated urban area in the United States. The island of Manhattan, 13 miles long and 2.3 miles across at its widest, holds just 7 percent of the city's land area but nearly 20 percent of its total population of 8 million people. Not surprisingly, the country's most populous city has one of the most complex transportation systems, featuring nearly 23,000 center-line miles of roads, streets, and highways; approximately 500 route-miles of commuter rail; and 225 route-miles of rail rapid transit. The city is served by three major airports, and by the largest maritime facilities for passengers and cargo on the East Coast.

Given the difficulties of traveling by car in such a congested area, millions of commuters, visitors, and residents alike rely on public transportation. On weekdays in 2001, New York City Transit subways and buses—the largest such systems in the country—served 6.4 million passengers. Meanwhile, weekday ridership on PATH trains and aboard public and private ferries totaled 258,000 and 91,600, respectively.

Overseeing the bulk of this extensive transit network were the Port Authority of New York and New Jersey, the New York City Department of Transportation (DOT), and the Metropolitan Transportation Authority (MTA), which ran New York City Transit. At the first sign of trouble at the World Trade Center, officials and operators of these various transportation agencies acted almost instantaneously to protect passengers, personnel, and equipment.

One minute after the North Tower was hit, an MTA subway operator, who had pulled his train into Cortlandt Station, alerted the MTA Subway Control Center to an explosion at the World Trade Center, initiating emergency procedures. Simultaneously, a quick-thinking PATH train dispatcher ordered the PATH trains that were en route at 8:46 A.M. to speed through the trade center station and head back out without opening their doors. Within minutes PATH officials shut down services, rerouting or canceling trains. These steps kept passengers from entering the danger zone, but they hindered people's efforts to exit as well. The split-second decisions that many made that morning ended up deciding their fates.

CHAPTER 2

"*Shut it down! Shut it down!*"

KENNETH SUMMERS'S CHOICE TO HEAD TO THE POST OFFICE just before 8:45 A.M. charted the course for his future. After leaving his desk at the Empire Blue Cross Blue Shield offices on the south side of the twenty-seventh floor of 1 World Trade Center, he exited the building through a revolving glass door and noticed a coworker running, ducking as he ran, indicating to Summers to look up. Seeing debris falling, Summers bolted back into the building to avoid being hit. From inside the revolving door he noticed the air in the lobby suddenly change color. Then he felt heat.

Shattered glass flew at Summers's face and he found himself lying outside next to a planter with black smoke and burning rubble all around. When Flight 11 struck the upper floors of the tower, some 10,000 gallons of jet fuel ignited into fireballs—one as big as 200 feet wide. The explosions shot down elevator shafts into the lobby, blackening an expanse of wall and blowing out windows. Summers slapped out the flames on his hair and shirt. In shock, and covered with burns, he walked to the corner of Liberty and West Streets and sat down on some steps.

While Summers was attempting to leave Tower One, Bonnie Aldinger had been making her way toward Tower Two. In the two months since she'd been laid off from her technology officer job at Fiduciary Trust Company International on the ninety-seventh floor of the South Tower, or 2 World Trade Center, Aldinger had been spending every possible moment paddling her kayak around New York harbor. On this warm Tuesday morning, however, it was back to business as she headed to a company-sponsored outplacement workshop on the ninety-third floor. She'd taken the subway, exiting the train in the World Trade Center Concourse, a combination shopping center and transit hub.

Boasting more than 60 stores, services, and restaurants comfortably accessible in any weather, Manhattan's largest indoor shopping mall stretched for two blocks beneath the Austin J. Tobin Plaza. Establishments like Fanny Farmer Candy, Chandler Shoes, Optical World, and Tower Coffee Express drew customers into the vast subterranean shopping center, and the three subway platforms connected through long passageways served approximately 150,000 people a day.

At 8:46 A.M., Aldinger had just placed her palm against the door to exit the concourse when a piercing screech ripped through the sky, followed by an earth-shaking boom. The brim of her ball cap shielded her overhead view, but she could see the faces of people in the plaza as they looked up, screamed, then bolted toward the doors where she stood. A swarm of people stormed in her direction, and Aldinger didn't stick around to see what had made them run. Instead she shot back underground, careening past a newsstand, down a set of stairs, and around a corner to get out of the path of any wreckage that might fly through the doors. She ducked into a restaurant where patrons huddled against a back wall as far from the entrance as possible.

Restaurant workers gave her some water while she waited for the crush of people bolting through the corridors to wane. She figured she'd continue on to her meeting once things settled down a bit. "This being New York, a small plane crashing low on World Trade Center One would not necessarily preclude attending a workshop on the ninety-third floor of World Trade Center Two," she later wrote to friends and family. "There would probably be a delay while emergency personnel did their jobs, but life would go on."

But by the time she stepped from the restaurant, police had started evacuating the mall, and the doors to the street were no longer an option. So Aldinger headed toward the subway station, walking north through an underground corridor. She warned the commuters she passed to turn around. She figured she'd make her way north and then climb to street level to find another way to get to Tower Two. She was still underground when the second blast hit, reverberating like a bass drum. "You felt it as much as heard it." Aldinger was swept up in the stampede, already running when, for the first time in her life, it occurred to her that she might be about to die.

"The plane that crashed just blew up," she heard someone say. But, to her, the blast sounded too big for that. She considered the possibility of a bomb. She worried there might be more. While debating whether she'd be safer above or below ground, she heard a train pull into the station. Anyplace that train was headed had to be better than here, so she slid her MetroCard through the card reader, pushed through the turnstile, and sprinted toward the open car doors. A woman stood on the platform looking confused. "Just get on the train!" Aldinger shouted. The other passengers stared, but she was too distressed to explain.

Just as Aldinger had been trying to exit the trade center concourse, Gina LaPlaca had been entering. Shortly after 8:30 A.M., the 22-year-old beginning law student had left her apartment in Gateway Plaza, an apartment building overlooking North Cove, the small harbor due west of the trade center towers, on the Hudson's eastern shore. Rushing to the subway, she'd accepted that unless the A train arrived right when she reached the platform she was going to be "embarrassingly late" for her nine o'clock Property class at Fordham University Law School. She'd moved to New York City only four weeks earlier. And after staying up late with a college friend who was in town for some job interviews, LaPlaca had snoozed her alarm a few too many times.

Now, with a messenger bag full of heavy textbooks slung across her body, she exited South Bridge, the pedestrian walkway that crossed over West Street (the multilane highway running up Manhattan's West Side that bordered the trade center). Crossing in front of the open door of FDNY's "Ten House" on Liberty Street, LaPlaca caught the eyes of two firefighters standing outside. She felt their gaze as she passed in her knee-length skirt and platform sandals. The flirtation made her smile a little, even in her hurry, and the loud, fast house music thrumming through her headphones matched her pace as she entered the World Trade Center Concourse on her way underground.

LaPlaca had just passed a newsstand, noticing the flower bouquets standing upright in their buckets, when a huge boom erupted, so loud that it was audible above the throb of the Paul van Dyk track coming through her earbuds. Suddenly all the people who'd had their backs to her as they, too, had entered the concourse now seemed to spin around in unison to bolt toward the door. Years later, LaPlaca would describe the massive about-face as like stumbling into the middle of some sort of flash mob. The choreographic effect was all the more disorienting given the dance music pounding in LaPlaca's ears.

In actuality, the source of the sound was unclear, so while some people sought safety inside the concourse, other people raced to get out. All LaPlaca could process in the moment was that something bad was happening. *A maniac with a gun must be shooting*, thought the self-described "suburban Jersey girl." *If a bunch of New Yorkers are running, there must be a good reason*. She joined the sprint toward the exit.

LaPlaca didn't stop to look up as she burst through the doors. Instead she kept running across Church Street. A block and a half later she finally turned around to see smoke, glass, and debris spilling from 1 World Trade Center. Papers fluttered through the air and a murmur traveled through the gathering crowd that a plane had hit the tower. A vendor in a boxy silver bagel and donut cart had a radio, and LaPlaca joined a small group of people clustering around to listen. "It was odd the way information flowed," she explained. "The closer you were to it, the less you knew what happened."

Her cell phone rang. LaPlaca's mother, who had heard the news in her office in New Jersey, was relieved to reach her daughter, and now urged her to head uptown to class at Fordham Law School. But LaPlaca's first priority was to find her visiting friend. One of the friend's interviews, she knew, had been scheduled for 9, or 9:15, in one of the twin towers. *Had she left the apartment? Did she make it to the trade center? Is she wandering around somewhere?* LaPlaca kept dialing her schoolmate's cell, but the calls wouldn't go through. The crowd gathering on Church Street grew and soon police pushed everyone farther east, into Zuccotti Park, preventing anyone from heading back toward the trade center site.

Before long the narrow park was so packed it was hard to move. LaPlaca was staring up at the burning tower, trying to strategize a plan when she heard, shooting in from the south, a whoosh like a missile. "It came almost too fast to see that it was a plane," she recalled. "We saw this thing go swooping into the building. That's when the panic ensued." The mass of people surrounding her surged back, recoiling as a fireball burst through the side of the building.

Back in Gateway Plaza, a neighbor unknown to LaPlaca named Jerry Grandinetti had just flopped onto his couch and clicked on the Weather Channel when the first plane hit. The huge boom drew him to the windows of his nineteenth-floor apartment overlooking North Cove. Above and over to the right, he saw thick black smoke peel around Tower One from the north face of the building. Reflexively, Grandinetti reached for the 35-millimeter camera that lay on the windowsill. He'd left it there a few days earlier after snapping aerials of his charges, the three dinner boats moored in the marina below that he captained for VIP Yacht Cruises.

Grandinetti kept clicking the shutter as smoke pouring off the engulfed tower turned from light gray to dark gray to black. At first white tendrils wisped out through broken windows in the building's western face. Then the whole top of the building was obscured. Wind blowing from the northwest fed the fire through the airplane-shaped hole on the north side, dragging a column of smoke south and east. Soon walls of crimson flame several stories high raged through the tower's fractured south face.

"It started out small and got rolling big," Grandinetti recalled. "It got rolling bigger and bigger." *This is real bad*, he thought.

An instant later he heard a piercing hiss like a rocket slicing through the sky. A volcano of fire and coal-black smoke erupted out the side of the South Tower sending a cascade of debris raining down. *We're under attack*, thought Grandinetti, as did so many others at that very same instant. The plaza around North Cove swarmed with people fleeing. Though Grandinetti was hanging out a nineteenth-floor window fewer than 300 yards from the burning buildings, no sense of immediate danger penetrated. "I was just kinda mesmerized watching stuff fall, watching the second building catch on fire like the first."

He had his camera to his eye when a man plunged out of the tower. "Aw, fuck!" Grandinetti shouted, loudly enough that a police officer in the parking lot below heard him and looked up.

"You've gotta get out of here!" the cop yelled up at him. "You've gotta evacuate!"

Still in his boxers, Grandinetti yanked on a pair of shorts and a T-shirt, stuffed his camera and some binoculars into a bag, and left his apartment. But he didn't evacuate the area. Instead he went to check on the three dinner boats he captained—*Excalibur*, *Royal Princess*, and *Lexington*—berthed along the northwest edge of North Cove. Discovering soot but no damage, he looped back around the cove toward the VIP Yacht Cruises office on the Gateway Plaza building's first floor, confronting along the way a direct view inside the gashed North Tower.

Through binoculars he could see, up close, the faces of people waving their arms and crying out, standing in the open air where once there was building between them and the sky. One woman's face, in particular, still haunts him. He remembers seeing her flailing, frantic, trapped below the fire. He could read her lips. "Help us! Help us! Help us!" she screamed. "It was cherry red, burning fierce. Cherry, cherry red." *This is fucked up*, he thought. *Why aren't helicopters rescuing these people?*

The simple answer to Grandinetti's question was that the Port Authority's plan for escaping fire in the towers did not include roof rescues. Mass evacuations by helicopter were never considered a viable option, although no signs in stairwells or statements made during regular fire drills announced that policy. Further complicating matters was that helicopters *had* actually rescued several small groups of people from the roof following the 1993 bombing, and news stories publicizing these rescues had reinforced the idea that the roof offered a path to safety.

Unlike in 1993, however, thick smoke precluded helicopters from offering a way out. "We're going to be unable to land on the roof due to the heavy smoke condition at this time," radioed NYPD Detective Timothy Hayes at 8:58 A.M. His chopper, Aviation 14, was the first police helicopter to arrive after the first plane hit. Even as NYPD high-rise experts prepared for the possibility of rappelling, the chief of the department forbade the Aviation Unit from attempting any heroics. As the morning progressed, conditions would only grow worse. Yet because 911 operators and FDNY dispatchers did not know roof rescues weren't possible, they weren't able to advise callers about the dangers of attempting to reach the roof rather than making their way down toward their only possibility for safety.

At 9:03 A.M., sailing yacht *Ventura* Captain Patrick Harris, the man who had called in the first strike to the Coast Guard, was standing on Church Street, at the eastern edge of the World Trade Center complex, gaping at the gash in the North Tower. After radioing that initial report, he had grabbed his handheld VHF radio, stepped off the *Ventura*, and crossed the World Financial Center plaza to get a closer look at the wound, a cookie-cutter shape of the plane's fuselage and wings. When the second jetliner swooped in low, pivoting at the last second, then slammed into the South Tower, he couldn't hear the explosion over screams from the crowd around him.

The instant the plane hit, people began fleeing north. Harris ran against the flow, dodging people on his way back to the boat that was both his home and his livelihood. He bolted down Vesey Street, crossed the West Side Highway, and entered the Winter Garden Atrium, a 10-story, glass-domed pavilion that offered the shortest route to his boat. He hadn't anticipated sheets of glass falling and shattering around him as he sprinted to where the atrium opened out onto the plaza around North Cove.

At last Harris clambered aboard the *Ventura*. He was preparing to pull the yacht away from the slip when he spotted his first mate, Josh Hammitt, about 100 feet away. Hammitt was rushing toward the docks, hand-in-hand with his very pregnant wife. They had been at the base of the South Tower when the second plane hit. Glass and debris had rained down on them. "Where's my family? I've gotta get my family," Hammitt shouted, panicked. His brothers and sister lived in Gateway Plaza, the apartment complex where Grandinetti had been hanging out the window. As if on cue, Hammitt's siblings, having fled their apartment to the nearest outdoor space, appeared in the plaza.

"Go get them," said Harris. "Get them on the boat and we're going to pull it out of here."

Then something caught Harris's eye—a peculiar movement in the air. Amid the debris spilling forth from the buildings, he saw what he came to recognize as humans in the sky. People were plunging from the towers. Some fell straight down. Others appeared lighter. Harris noticed how one woman dropped until the wind caught her, flipped her around, and pushed her back up briefly before letting her plummet once more. He tried to push the image from his mind so he could focus on readying the boat for departure.

Soon Hammitt returned with two brothers, his sister, and their dog. Harris tried to ease his mate's panic by helping him click into the familiarity of routine. "We're going to pull out of here just like we do with a charter," Harris said. "We're going to do it by the numbers." The captain called out orders as he throttled forward on the spring line, cast off, and backed the boat from its berth.

Once he'd made it safely out of North Cove, Harris paused at the helm, trying to figure out what to do with his passengers. Then he noticed that, below her billowy maternity dress, the mate's wife had blood on her legs. *She's gonna have the baby on the boat!* Harris panicked. He spotted an NYPD boat nearby, figuring the crew would know what to do. But a moment later, he registered that what he saw wasn't her own blood but "splatter" from falling bodies, from when she had walked through the plaza at the foot of the towers. "I looked closer and there was actually somebody's index finger plastered to her calf." Without her being aware, he said, he "discreetly pulled the finger off and threw it in the water."

Tens of thousands of emergency workers put themselves in danger that day—and as the disaster unfolded, civilians, too, found themselves thrust into the role of first responders. Rich Varela was one such person. A few minutes into his workday, Varela looked around the windowless, nearly soundproof "comp data" room humming with servers, telephone switchboards, and other electronics and noticed that things seemed oddly quiet. Usually by this time in the morning the phones would light up—a luminous indication of the frenzy ramping up on the company's trading floor down the hall. *There must be a late bell today*, he reasoned, continuing his work backing up hard drives that contained months of recorded phone transactions. A telecommunications specialist for an information technology firm, Varela was working a contract gig for a company on the twelfth floor of 1 World Financial Center, directly across the street from the South Tower.

About 15 minutes after he'd sat down at his desk, a "crazy, ridiculous rumble" erupted, and the building did a little shimmy. *Maybe it was one of those big 18-wheelers*, thought Varela, picturing the thunder created when a large truck rolls over metal plates in the street. The shake of the building was enough to spur a passing thought about his isolation inside his computer-room tomb. *If that* had *been an explosion, from a blown gas line or something, nobody would even know I was in here.*

A few minutes later, his cell phone rang. "Rich, where are ya?" asked a buddy calling from Jersey.

"I'm at work."

"Get outta there! They're crashing planes into the World Trade Center!"

"What are you talking about? I just came out of the World Trade Center." Less than a half hour earlier Varela had, indeed, exited the PATH train at the World Financial Center stop, climbed the escalator leading into the South Tower, and crossed over West Street through South Bridge, the covered pedestrian walkway, before heading to the twelfth floor. He had been running a good 15 minutes behind schedule when he'd passed through 1 World

Financial Center's glass-encased lobby and stepped into an elevator. At about 8:47 A.M. when the doors opened on his floor, he'd walked 20 feet down the hall into the cloistered chamber of the comp room and logged into a desktop terminal. He'd seen nothing, talked to no one before sealing himself into the comp room.

Because his friend had a history as a prankster, Varela didn't buy his story right away. "No, I'm serious," the friend said. And then it clicked. That rumbling sound. Varela gathered his things, neither panicked nor dawdling. He opened the comp room door and took a few steps down the hall. In a nearby office, phone receivers dangled off corkscrewed cords, papers lay strewn across worktables, and chairs loitered at odd angles rather than nestling neatly under their desks. The trading floor was wholly uninhabited. "It was like people just evaporated."

The building's fire alarm screeched and strobes flashed near the ceiling. A security guard at the end of the hallway spotted him. "What are you doing in here? Get out of here!" Varela figured it was safest to take the stairs, but just as he approached the stairwell an elevator opened in front of him. *Shit. Should I take it? It's only 12 floors...* He jumped in.

After receiving the call from the Coast Guard's acting commander, Captain Harris, Rear Admiral Bennis had pulled onto the shoulder of I-95, grabbed a battery-powered television from the trunk of his car, and handed it to his wife. "Oh, here's an instant replay," she said, thinking the plane on-screen was the first. But no. The time was 9:03 A.M. The footage was live. The second plane, United Airlines Flight 175, had just plowed through the South Tower. The rear admiral turned the car around, gunning it back toward New York.

As acting commander, Harris knew that it was his job to coordinate the Coast Guard's response to this unprecedented emergency. To fulfill the Coast Guard's duty to secure the harbors of the homeland, he would ultimately enforce a closure of the port to all nonessential traffic and dispatch a fleet of vessels to patrol the area. After that, the mission would be far from clear. There was no plan book for responding to a commercial airliner attack on the World Trade Center. None of the other agencies with a presence on the water had any plan for this either.

On an ordinary day, any number of different organizations, municipal or military, could be called upon to respond to an emergency in the Port of New York and New Jersey, largely determined by which agency received the call first. While it may seem odd to the layperson, the operations of the Coast Guard, NYPD Harbor Unit, FDNY Marine Division, New Jersey's police and fire marine units, and other authorities ran along rarely intersecting, predominantly parallel tracks.

Occasionally a mutual aid call brought the responders together (if the Coast Guard requested diver assistance from the NYPD, for instance). The organizations also collaborated during annual drills where some combination of agency personnel came together to exercise emergency scenarios. These drills, which emphasized resource allocation and sharing expertise rather than establishing hierarchies, helped to foster the harbor's collaborative ethos.

In the face of this massive incident, the shared purpose and common ties that connect mariners of all types ruled the day as the different agencies cooperated with civilian boat crews. As it turned out, the lack of a plan wound up setting the stage for creative problem solving and improvisation. Throughout this historic morning, the New York harbor community joined forces to carry out an unprecedented and remarkably successful evacuation effort.

Carlos Perez was out on the water, approaching the southern tip of Governors Island, when he spotted the second jet flying so low over the Hudson that "it looked as if it were coming in for a landing." Just after the first plane hit, the boatswain's mate third class—a Coast Guard designation for a master seaman expert in navigation, small boat operations, search and rescue, small arms, and other duties—had set out from Fort Wadsworth aboard the utility boat *41497*. That day Perez's unit, including an engineer and a crewman, had been assigned to the first search and rescue "ready boat," charged with standing by prepared to launch at a moment's notice, day or night. But when Perez first heard the search and rescue alert, he shrugged it off, knowing that the alarm had been malfunctioning lately. Not until he'd stepped outside and spotted smoke pouring from the North Tower did he process that this emergency was real. In accordance with the Coast Guard's motto—*Semper Paratus*: Always Ready—Perez and his crewmates got under way and shot across the harbor toward Lower Manhattan.

Now, about 10 minutes into their run, the second plane appeared, vanished, then reappeared dead ahead. "It was as if it dropped down directly in front of us." Perez watched the jet streak toward Manhattan. The engines revved faster and louder. Then the plane banked swiftly to the left.

He and his crew stood dumbstruck. Perez called out the order to keep a sharp lookout, 360 degrees—in the air, in the water, everywhere. He received word from his command that the city was under terrorist attack and that additional aircraft were currently unaccounted for. His orders were to patrol the area around the southern tip of Manhattan to rescue anyone in the water and recover any debris that might be used as evidence.

Through binoculars, Perez watched people attempting to escape the unendurable conditions inside the tower by trying to scale the side of the building. The glasses revealed "horrific silhouette images" of people jumping.

"The worst feeling as a first responder in any capacity," explained Perez, "is being on a scene of distress and not being able to do anything about it." This sentiment, shared by countless mariners as they began mustering on Manhattan's shores, would become a driving force that propelled so many into action.

Coast Guard Lieutenant Kenneth "Bob" Post watched the second attack unfold on-screen. As a watch officer at the Vessel Traffic Service (VTS), his job was to supervise three stations monitoring vessel traffic in different areas of the Port of New York and New Jersey. With eyes glued to the feed from the camera that had been trained on the burning North Tower seconds after the first collision, Post suddenly noticed the wavering heat trail of exhaust from the second plane bearing down.

"Shut it down! Shut it down!" ordered Commander Daniel Ronan as he burst into the room moments later. He was referring to commercial operations in the port. Ronan had watched the fireball explode out of the South Tower, then raced the mile and a half to the VTS from his dentist's office near the Stapleton, Staten Island piers. As director of the waterways management division for Activities New York, Ronan was charged with ensuring that the thousands of commercial vessels transiting through New York harbor got to their destinations safely. The last thing he wanted, with all that was happening in Lower Manhattan, was tugboats or large cargo ships out on the waterways, attempting to conduct business as usual. By the authority vested in him by the Code of Federal Regulations, or CFRs (the federal laws governing shipping within the United States), through what's called a vessel traffic notice, he could declare an emergency and stop all maritime commerce. Despite the decision's serious financial ramifications, that was precisely what he did.

New York City had grown up around its waterways. Sheltered from the sea's storms and rarely icebound, the harbor offered natural deep-water channels and two separate river routes to the ocean. These features helped establish the Port of New York as the busiest in the world from 1830 to 1960. In its heyday, Manhattan's 43 miles of direct waterfront had been developed, with finger piers lining the entire western shore of the island, into 76 miles of usable frontage. Later, changes in shipping that required far more storage and machinery space than the narrow island could accommodate pushed commercial port activities off Manhattan's shores, into Brooklyn and New Jersey.

The major shipping advancement that ultimately bumped the port into third place nationwide was actually launched here in 1956, when an aging tanker, the *Ideal-X*, was converted into a makeshift container ship—the first of its kind—and transported 58 aluminum truck bodies to Houston, Texas, where 58 trucks waited to drive the metal boxes to their destinations. The birth of containerization—shipping cargo in standardized crates loaded whole from ship to shore—spelled the end of break-bulk cargo, where items were loaded and offloaded to and from ships' cargo holds by hand. It also spelled the end of the old working waterfront.

By 2001, nearly all the remaining traces of the industrial harbor had vanished from Manhattan's shores. The Port of New York (and subsequently) New Jersey, which included a system of navigable waterways along 650 miles of shoreline, was still alive with commercial activity. Now, though, it happened largely out of view.

Commander Ronan had recently participated in a Port Authority exercise with representatives from global shippers, major banks and investment companies, and the auto industry, among other relevant parties, to ascertain the economic impact of supply-chain disruptions caused by closing the port. At the Port Authority's five container facilities alone, the economic impact tally totaled $20.5 million per day. What, then, would the closure of the whole port cost each day? Ronan knew the answer: $1 billion. Still, he proceeded. "We didn't know what was going to happen next," he explained. "We weren't sure if a ship coming in was going to be the next wave."

"Reach out to the Sandy Hook Pilots," he told his watch officer. "Let's not bring any ships in." By calling the Sandy Hook Pilots—the licensed maritime pilots that boarded every passenger liner, tanker, freighter, or other oceangoing vessel and guided its navigation through the harbor—the Coast Guard was alerting the gatekeepers that the port was now closed. Then he broadcast the announcement over marine radio: "The Port of New York/New Jersey is closed to all commercial traffic. No movements are authorized by direction of the director of the Vessel Traffic Service of New York."

Between 300 and 400 vessels were likely moving through the port at this time. The shutdown, as Watch Officer Post understood it, meant "everybody stops, everybody goes to a berth, everybody goes to anchorage, nobody comes in and nobody goes out. Our job became securing everything." Radio traffic exploded. The VTS team was besieged with calls as they struggled to guide the pilots, mates, and masters of vessels of all shapes, sizes, and configurations toward safe berthing and mooring locations throughout the harbor.

"Everybody's screaming on the radio," recalled Post. As each vessel appeared on radar, the watch standers could click its name for size, draft, and cargo information that helped determine where the ship could safely go. Each monitor had an electronic navigation chart overlay to map out specific areas that watch standers used to identify secure options. "You're telling them go to this pier, go to this anchorage, proceed out and continue on. Everybody on the outside coming in we told not to proceed. Of course they had to proceed because you can't just turn a friggin' 900-foot vessel around very easily."

As the Coast Guard shut down the port to commercial traffic, other maritime forces kicked into action. In a firehouse on Pier 53, two miles up the Hudson River, a voice alarm reporting an explosion at the World Trade Center brought the whole marine fire company to their feet. Marine engineer Gulmar Parga stepped outside to hear people on the adjacent pier yelling to them that a plane had hit one of the towers. There was no question that this FDNY company, Marine One, would make this run. What was unclear was how they'd be able to assist once they arrived, given that the shoreline was 333 yards away from the foot of the South Tower. *We're a fireboat,* Parga recalled thinking. *What are we gonna do at the World Trade Center?* He did not anticipate the critical contributions that New York City's fireboats would ultimately make.

Within minutes Parga had engines running aboard the 129-foot, 335-gross-ton *John D. McKean,* and the boat was southbound on the Hudson, headed for the World Financial Center. At 8:59 A.M., Captain Ed Metcalf (who was covering a shift, working only his second day with the company) called in to report the view from the river: "This is Marine One. We're in the river. You've got fire out of the north side and now coming out of the west side of the World Trade Center. The west side. . . . Fire has penetrated the skin."

Parga's first thought when he caught a glimpse of the North Tower was: *A lot of people are going to die today*. To seasoned firefighters, the black smoke rolling off the upper floors of a supertall skyscraper warranted serious apprehension. It meant the fire was burning freely, with no water reining in the blaze.

Soon pilot Jim Campanelli was rounding up to dock on the southwest corner of North Cove at the foot of Liberty Street, fewer than 1,000 feet west of the South Tower. With no proper bollards to which the deck crew could make fast lines, they slipped the mooring ropes under the railing that topped the seawall and tied them to trees. This deprived the *McKean* crew of the ability to swiftly let go lines should conditions warrant a quick exit, but they didn't have much choice.

Captain Metcalf immediately left the boat and headed inland to receive orders from command. Until he returned, the *McKean* crew would be left in limbo, looking up at the smoke and flames and watching bodies fall. "You could hear them smacking into the ground," Parga explained. "You could see groups of three jumping at the same time, all holding hands. . . . Big booms. . . . They were falling like leaves from a tree."

In the waterfront plaza surrounding North Cove, people who had been enjoying a sunny day, riding bicycles, walking dogs, and pushing strollers, now darted about, unsure of what to do next. A slow stream of injured began making their way to the river's edge. *McKean* firefighter Billy Gillman rushed to the aid of a man, sitting him on a bench across from the boat. "The whole top of his head was burned and the skin was coming off," Parga recalled. The man was Kenneth Summers.

After getting caught in the fireball that exploded into the lobby of the North Tower, Summers was resting on a set of stairs on the southwest corner of Liberty and West Streets when he heard the whine of Flight 175. Flames erupted out of the side of the South

Tower in his direction and soon he was running, pulled along by the crowd, away from the tower and toward the river. His serious, very visible injuries drew the attention of a man who offered to help Summers get medical attention, explaining that he was the son of a doctor. A woman offered her phone. After leaving a message for his wife, Summers heeded the man's advice to head for the ferries bound for New Jersey. Before long they spotted fireboat *McKean*, tied up along the seawall.

"We got one!" hollered Gillman as he jumped off the boat to help, eager to offer whatever assistance he could. The firefighters placed an oxygen mask over Summers's burned face and gave him water. "They also tried to clean me up a little with a wet cloth but stopped when they realized that it might be too painful," Summers recalled. He clearly needed more specialized medical care than the firefighters could provide, so soon he and the doctor's son continued on to the ferry.

At the World Financial Center ferry terminal, the crowd "parted like the Red Sea" when Summers approached. "There were thousands of people trying to escape Manhattan and still everyone cleared the way," he recalled. "There was no panic. They let us pass right up to the edge of the ferry dock. The ferry had just pulled away from the dock and when they saw me standing there, they nosed back in." Almost immediately, from the moment the first plane hit, commuter ferries had become waterborne ambulances.

The crew of FDNY's Marine 1 remained in limbo, waiting for word from Captain Ed Metcalf. The radio channels were clogged with overlapping, often indecipherable transmissions, making it impossible for the *McKean* crew to reach him.

In the aftermath of the 1993 bombing, the FDNY had learned how ineffective their radios were in the difficult high-rise environment of the twin towers. First, the weak signals often failed to penetrate the steel and concrete floors that separated fire companies who were attempting to communicate. Second, having so many companies using the same point-to-point channel left many transmissions unintelligible. Reliable communications in skyscrapers required an amplifier to boost a radio's signal so it could reach a building's highest floors. Although signal-boosting repeater systems had been installed in the towers in 1994, the chiefs on scene could not make them function properly on that crucial morning seven years later.

In September 2001, the FDNY carried analog, point-to-point radios. Companies responding to a fire typically communicated on a single tactical channel that was monitored by chiefs on the scene. Those chiefs would also use a separate command channel as well. At the World Trade Center that morning, the constraints of the equipment caused deadly failures to communicate among firefighters and their command. The higher the units climbed in the towers, the worse the transmission troubles became. Chiefs in the lobby often received no response to their attempts to reach particular units higher in the buildings. Even after the South Tower collapsed, many firefighters in the North Tower did not hear the order to evacuate. Tactical 1 had never been designed for use by so many people at once. Even when radios could receive transmissions in the high rises, other communications often drowned out evacuation instructions that could have saved lives.

Instead of further clogging the radio channels, the *McKean* crew had decided to lay out tools and gather up emergency supplies to continue helping the people arriving on the waterfront, burnt and bleeding. The sense of helplessness in the midst of the mayhem soon became unbearable. "I wanted to run up there and start dragging hose and doing whatever it was I could do," said Parga. "We wanted to make ourselves useful." But other crew members argued that they shouldn't stretch supply lines in this spot in case they were called to duty elsewhere. "We were really going at each other—fighting amongst each other because anxiety was building," Parga recalled. "We were yelling."

The argument went full circle, bringing them right back to where they started: they needed orders from the captain. Parga donned a set of firefighting "turnout" gear he found on board that almost fit him. (He had left his own helmet, pants, coat, and boots at Marine 9 on Staten Island after working a tour there, thinking: *I'm an engineer. When am I gonna use bunker gear?*) Once changed, he announced his intention to go inland to find Metcalf. But another engineer protested, saying, "You'd be deserting your post if you go up there!"

More yelling ensued until firefighter Tom Sullivan, who was technically off duty, offered to go. Because the attacks had occurred so close to the nine o'clock shift change, many firefighters just going off duty had responded to the call. Some had self-dispatched while others had been given permission by company officers to "ride heavy" and join on-duty firefighters headed to the scene.

As an extra guy, Sullivan got off the boat, crossed the plaza, and started walking east on Liberty Street, in hopes of locating the captain. He was heading directly toward the towers, both of which were burning furiously but still standing. It was a few minutes before ten o'clock.

CHAPTER 3

"NEW YORK CITY CLOSED TO ALL TRAFFIC"

WHEN TELECOMMUNICATIONS SPECIALIST Rich Varela stepped out of the elevator in the lobby of 1 World Financial Center shortly after nine, he caught his first glimpse of what looked like Armageddon. The whole front of the building had imploded leaving the lobby's plate glass windows in shards amidst the smashed concrete and polished stone. Pockets of flame feasted on combustibles.

Earlier that morning, when Varela had first arrived, he must have stepped into the elevator right as the first plane hit, the shaft shielding him from the noise of the impact. He hadn't lingered in the hallway on floor 12, and once the comp room door had closed behind him, Varela was sheltered from the commotion that had prompted the mass exodus from the building. Now he found himself alone amid the rubble.

The first thing Varela saw as he stepped through a doorway that left him standing beneath the pedestrian overpass that he'd crossed on his way into work was a man in a business suit, half burned. "He was roasted on one side, just kind of walking around aimlessly," recalled Varela. "It looked like something out of a zombie apocalypse." The man in the suit may well have been Kenneth Summers, still dazed after being caught by the fireball in the lobby.

Varela looked across West Street and saw the blazing hole in the side of the South Tower. He stood there staring into the stillness, into the silence. "You could hear a pin drop. You're in Lower

Manhattan. You could hear sirens in the distance but immediately in the area there was no motion of life. I thought that was so eerie." Slowly a handful of office workers gathered nearby.

The congregation stood gaping, in shock, until a security guard barked at them. "Get the fuck out of here. It's not safe. You gotta get out of here," Varela recalled him saying. "He's going crazy about us leaving that area."

Now the group of office workers that had swelled to about 10 began walking toward the water. Varela thought about the architects who'd built the towers and how remarkable it was that the buildings hadn't fallen. Indeed, the towers had, so far, withstood the unthinkable—the unthinkable that had actually, in fact, been thought of decades before.

Even a Boeing 707 couldn't knock the buildings down. Such had been the Port Authority's response to an ad run in *The New York Times* in May 1968. Paid for by an opponent of the twin towers' construction (the owner of the Empire State Building), the eerily prophetic ad depicted an airplane striking the north face of the North Tower, warning that a plane was likely to fly, accidentally, into the buildings. The Port Authority's retort, bolstered by calculations made by structural engineers and early computer simulations, held that the towers could withstand an impact across seven floors from a Boeing 707, the largest airplane flying at the time. The sense of the buildings' invulnerability had been reinforced in 1993 when a bomb that the FBI described at the time as "the largest improvised explosive device" in the history of American crime had failed to fell the towers.

Now, here again, the external columns—the iconic pinstripes—kept the buildings standing, holding the weight of all 110 stories and transferring it down through the foundation to the bedrock below. The towers seemed impervious to the massive damage the jets had inflicted.

Varela was standing, contemplating architecture, when a series of artillery-like blasts that reminded him of hearing tanks firing rounds at Fort Dix made him duck for cover. "I heard what I thought at the time was cannon fire or missiles coming into

Manhattan." Boom! The blasts echoed through the streets. Boom! Varela wondered, *Are there battleships out in the water shooting planes out of the sky?*

He turned around to face the towers and realized he was hearing the sounds of bodies when they hit the ground. He watched dozens of people plunge to the street. One man on an upper floor dangled near the northwest corner of the South Tower, clinging to a ledge. "He was hanging on, hanging on, hanging on. I just remember thinking, *Don't let go!*" But then the man somersaulted backward and flipped through the air.

A woman in the street crumpled at the sight. "Why are they jumping?" she wailed.

"They're probably better off," lamented Varela. Instead of burning alive they had chosen the sky.

Rick Thornton, a Navy reservist in his eleventh year as a New York Waterway ferry captain, also saw the falling bodies. From the helm of the ferryboat *Henry Hudson*, he watched them—"first one here, then one or two from another angle"—through the airspace between the two towers. "That's when we realized how catastrophic this was."

At 9 A.M., less than 15 minutes after the first plane struck, Thornton had just offloaded his passengers at the midtown Manhattan dock at West Thirty-eighth Street and was returning to Weehawken, New Jersey, to pick up his next load of commuters. Although he was aware of trouble downtown, he was still conducting business as usual. Suddenly, a fireball exploded, not from the first tower, but out the side of 2 World Trade Center. Instantaneously Channel 69, New York Waterway's "house frequency," burst forth with the voices of captains shouting about how a second plane had hit. Thornton didn't dither.

"I just turned my boat around," he recalled, explaining that "pure instinct" took over. Thornton's deviation from the tightly timed drop-offs and pick-ups that are the core function of a ferryboat could easily be construed as a huge dereliction of duty. "The boats obviously belong to New York Waterway and every run is scheduled," Thornton explained. "You don't just go off on your own. You just don't do that." But a combination of experience and intuition dictated otherwise.

Although he was scheduled to make a run out of Weehawken at 9:10, he steered the ferry south. "I didn't call anybody on the radio, I didn't check in with anyone on board to say 'I'm going offline,' I just turned my boat around and I was heading down toward the World Financial Center." His experience running ferryboat evacuations following the 1993 bombing likely influenced his decision to change course. He remembered how, back then, all the company's assets were put to the task of getting people out of downtown Manhattan. "Maybe I'm five minutes ahead of what they're going to order me to do, or I'm just completely overstepping my bounds," he explained. "But my pure instinct was just to head downtown."

Ten minutes after gunning the 18-knot *Henry Hudson* downriver at full throttle, Thornton queued up with other ferries waiting off the World Financial Center terminal to pull into an open slip. The Hudson churned with wakes. The ferryboat bobbed around as Thornton stood in the wheelhouse, staring at the raging fires and struggling to stay focused. "You're trying to divide all your attention now," he said. "You've got your radios going. Your phone is ringing—your primitive flip-top cell phone. Your crew is talking a mile a minute. You're hearing all this chatter on the radio, both professional and people just talking all over each other about what they saw: Did you see this? Did you see that? We gotta go here. We gotta go there. It's like a complete sensory overload."

Next he heard an announcement that the ferries should steer clear of the World Financial Center terminal due to concerns about broken gas mains underground. So he headed south through the haze of smoke that the north wind pushed toward the Battery. The impact of the second plane had made clear to everyone—emergency responders, maritime workers, and civilians—that this was a coordinated attack. No one knew what would come next. There was no more business as usual.

All along the seawall, throngs of people, panicked, desperate, and trapped, pressed against the railings. Behind them the blazing towers belched black smoke into the sky, and before them churning currents roiled the waters of New York harbor. Not everyone knew how treacherous those currents could be. People wanted off Manhattan Island. And some decided to jump.

Suddenly one of Waterway's high-speed catamarans shot past Thornton's ferryboat. The catamaran's captain announced over the radio that he was racing to rescue a swimmer caught in a ripping current. By the time the boat reached him, Thornton recalled, the man had been pulled halfway to Governor's Island across a central harbor thoroughfare. Moments later, Thornton spotted another man in the water, right along the shoreline, "being dragged under the pier, under the pilings" with no way to climb out. The water yanked him along and, with nothing but slimy algae-covered pilings to cling to, the man seemed to be losing his fight.

Thornton cut the *Henry Hudson*'s engines and instructed his deck crew to set up the man-overboard ladder. Then he lined up for a slow approach to the seawall. As he nosed the boat in so deckhands could retrieve the drenched and exhausted man, people swarmed over the curved steel railing atop the seawall and onto the boat. "That's when people just started [climbing over]. They wanted to be rescued. They were not going to be denied. . . . It was like being the last lifeboat on the *Titanic*."

From the helm of a vessel certified to carry 399 passengers, Thornton surveyed the crowd before him. "They were smoke covered, bloody, broken arms and legs, children alone with no parents . . . people as far as the eye could see. . . . We had no idea what was going to happen next."

Among the people Thornton recalled seeing on shore was a blind woman, maybe in her fifties, with whitish hair, clutching a German shepherd seeing-eye dog. "She was just clinging to that dog so tight," said Thornton. Four men in business suits lifted her up "like a surfboard" and passed her over the railing to ferry crew members. *Does she have any idea she's boarding a boat?* he wondered.

Within minutes the ferry was packed, but people continued to pour over the wall. Thornton got on the loudspeaker: "Don't worry. We have more boats. We're coming back. We're coming right back." But his attempts at reassurance didn't stop people from jumping, even as Thornton backed the ferry away. "They were literally swarming on to the boat," he explained. "There was some measure of blind panic when we backed away from the pier." Several launched themselves across two, three, four feet of open water and ended up dangling off the bow railings until deck crews could pull them aboard.

Instead of managing vessel construction in some shipyard somewhere or bobbing about some godforsaken sea, Greg Hanchrow was, on September 11, 2001, a regular, landlubbing commuter. After many long years on boats—running tugs, working the decks of deep-sea research vessels, and steering small ferries and dinner boats—Hanchrow, for the first time, held a conventional, year-round, 40-hour-week job as operations director for Spirit Cruises.

His shift into a more settled lifestyle was well timed. Three months earlier he and his wife had adopted a two-and-a-half-year-old girl from Bulgaria. Today was his wife's first day back at work and their daughter Maeve's first in daycare. By 8:30 A.M., Hanchrow was making good time driving from his home in Rockland County, 15 miles northwest of Manhattan, toward Jersey City where he and his boss had a nine o'clock meeting at Liberty Landing Marina.

He was in the midst of a regular commuter pit stop, grabbing a coffee and a bite at Linwood Plaza in Fort Lee, New Jersey, when one of Spirit's captains phoned him about the first plane. The captain called again to report the second plane, and not long after, Hanchrow heard "all kinds of sirens in Fort Lee." Squad cars raced down the nearby Route 9W Extension to shut down the George Washington Bridge, which carried some 309,300 vehicles daily—more traffic than any other river crossing to Manhattan.

Within nine minutes after the second hijacked airliner struck Tower Two, the coordinated efforts of the NYPD, Port Authority, MTA, and DOT led to the quick closure of all the bridges and tunnels into Manhattan. New York and New Jersey police began shutting down interstates with major access points into New York City. By 9:12 A.M., variable message signs en route to the George Washington Bridge flashed a highly unlikely announcement for a Tuesday morning at rush hour: "Bridge Closed."

Bridge and tunnel closures meant all New Jersey Transit buses to and from New York City were canceled. Major area highways, including the West Side Highway, FDR Drive, parts of the Long Island Expressway, and portions of the New Jersey Turnpike, were closed to all but emergency vehicles. Police barricaded the entrance and exit ramps of local bridges and tunnels, and avenues in uptown Manhattan turned into high-speed lanes for police cars and ambulances shooting south toward the disaster. Soon digital signs across the whole metropolitan area announced: "NEW YORK CITY CLOSED TO ALL TRAFFIC."

Transportation shutdowns quickly ricocheted out beyond the New York area. Greyhound buses, part of the country's largest city-to-city bus line, were canceled throughout the Northeast, while all terminals within a one-mile radius of any federal building were shut down. Amtrak suspended nationwide service to allow for a comprehensive security sweep to inspect all trains, tracks, bridges, tunnels, and stations. Security personnel remained posted at all facilities, 24 hours per day, seven days per week, patrolling entrances and exits and restricting access. By 9:40 A.M., the FAA halted all flight operations at U.S. airports, for the first time in history. Three minutes later, a third plane, American Airlines Flight 77, crashed into the Pentagon in Washington, D.C. One minute after that, as the White House evacuated, commuter rail services into Manhattan, including the Long Island Railroad and MetroNorth, were interrupted or suspended.

I don't know what's going on, thought Hanchrow as he sat in his car in Fort Lee, hearing sirens all around. *But whatever's going on I gotta get the boats out of New York. I've got $20 million in assets with my name on them, and I gotta move them off the pier.* Radio reports confirmed that Manhattan was on lockdown. Hanchrow was stuck on the wrong side of the water, but he had an idea. He raced toward the parkway bound for the Petersen Boat Yard & Marina in Upper Nyack, New York, where he knew people—"salty old types"—from whom he hoped to beg a favor. At the time he was thinking only of moving the boats, but before long he would end up moving something much more irreplaceable: people.

Around the same time that Rick Thornton had rerouted his ferry toward downtown Manhattan, and Greg Hanchrow was scrambling to get across the Hudson to access his dinner boats, Tammy Wiggs, the newcomer to the stock exchange floor, was panicked but so far holding in place. Despite the growing sense of alarm, it wasn't clear what would happen next. "The stock exchange never closes," explained Wiggs, save for seven weekdays out of the year. "We didn't know what was going on, but we knew we couldn't go anywhere because we had to man our posts." Frantic communications erupted between stock exchange executives and city, state, and federal governments, as well as the U.S. Securities and Exchange Commission. Should they open the market? Should they evacuate? Could this be the next building hit?

When a crowd of people headed for an exit—one that Wiggs hadn't even known existed—she grabbed her purse and followed them onto the street. Outside Wiggs could see papers flying and white smoke billowing. A couple of blocks off Wall Street, she met up with several coworkers clustered on a corner. Everyone around her had worked in New York City longer, and all seemed to know what they planned to do next. But Wiggs had no plan. She didn't even know where she was.

"Can we leave?" she asked her new colleagues, over and over. "What if we open?"

"Don't worry about it," replied Merrill Lynch Director Karen Lacey. "Don't think that you have to stay. You're not going to be in trouble if you go home. Just go."

Another colleague chimed in with reinforcement: "If Karen tells you to go home, go home," he said.

But Wiggs was too frightened. She tried to call her older sister, who for the past four years had been working for Lehman Brothers in 3 World Financial Center, just west of the North Tower, but the call wouldn't go through. A few people invited her to join them in walking north along the east side, but Wiggs, who considers herself "directionally impaired," didn't realize how wide Manhattan is and feared winding up anywhere near the Empire State Building—which, in her mind, could be another potential target.

So when Lacey invited her young colleague to accompany her to Hoboken, New Jersey, Wiggs latched onto the idea. She didn't know anything about Hoboken or about Lacey, except that she was the senior woman on the floor, but Wiggs was relieved to let someone else take charge. "She kinda took me under her wing. She almost made the decision for me because I was just lost. I wasn't capable of making a decision at that point. For some reason I didn't think I wanted to go north, which ultimately would have saved me all this trouble."

At the time Wiggs didn't realize that Lacey's plan entailed moving west—toward, instead of away from, the burning towers.

Inside the towers, fires continued to rage. Noxious smoke poured through ventilation ducts, traveling freely between floors. Flames licked out through the jagged holes torn through the buildings' skins where soot-covered, stranded people clustered four and five deep, gasping for air. No one, not even the towers' engineers and architects, knew if the thin, lightweight trusses that made up the innovative flooring system could withstand a fire.

Remarkably, the fireproofing material—a mixture of mineral fibers and adhesive—that had been sprayed onto the webbed steel "bar joists" of two of the world's tallest buildings had never been tested. While the trusses, so much lighter than traditional beams and columns, had served the economics of the towers by allowing for more tenant space, there was no data about the floors' ability to survive a fire. Within 20 years of its application, the fireproofing was already failing and falling off. Shortly after the 1993 bombing, the Port Authority had begun replacing it. But by September 11, 2001, it had only finished about 30 out of 220 floors.

Unlike a typical office fire, fed with paper, rugs, desks, and drapes, the twin infernos blazing through multiple floors of both towers also feasted on tens of thousands of gallons of aviation fuel, which burned at much hotter temperatures, with pockets of fire reaching as high as 1,832 degrees Fahrenheit. This heat, according to the National Institute of Standards and Technology (NIST), "significantly weakened the floors and columns with dislodged fireproofing to the point where floors sagged and pulled inward on the perimeter columns."

In fact, by 9:50 A.M., the fires in the South Tower were burning so hot that the airplane began melting. Photographs analyzed months later by NIST revealed a stream of molten aluminum pouring from a window on the corner of the eightieth floor. About 50 minutes after the jetliner struck the building, observed *New York Times* reporters, "the eighty-third floor appeared to be draped across windows on the eighty-second floor, and was gradually drooping even lower."

Since just before 9 A.M., New York Waterway Operations Director Pete Johansen had been standing at the top of the ramp leading down to the World Financial Center ferry terminal, helping to manage passengers. Even after the second plane had hit, "We told everybody, 'Stay calm. We're getting you on the boats.' And everybody was fine with that," Johansen recalled. As one ferry backed away, another pulled into the slip.

On a typical day the ferry crews made an average of 68 dockings in an eight-hour shift and could fill a 400-passenger ferry in less than two minutes. From where they stood along the seawall, passengers could see how quickly each vessel arrived, filled up, and pulled away. That efficiency, Johansen imagined, eased apprehensions. What raised anxieties, however, were the fighter jets that swooped over the area shortly after the second plane hit. "My first reaction is, *Are those ours or are they theirs? Whoever 'they' are.* You hear these fighter jets racing overhead and you don't know what's going on."

At 9:13 A.M. the fighter jets had set a direct course for Manhattan, arriving at 9:25 A.M., 22 minutes after the second plane struck, to establish a combat air patrol over the city. These jets (indeed "ours") were the ones that spooked Johansen as he continued directing people down onto the barge in batches to avoid overcrowding.

Passengers arrived in waves, but even when lines grew long Johansen said he "didn't think of loading over capacity" because doing so might have invited further tragedy. "Think about if you flipped a boat how that would have changed things," he explained. By his account, the Waterway's deckhands kept up with their clicker counts without exceeding allowances.

Occasional lulls between boardings gave Johansen moments to reflect: "One of the things that goes through my mind is, *If the building were to tumble over—if you laid the building down—would it touch us?*" Technically speaking, the answer was yes. The North Tower stood 1,368 feet in the air and Johansen was working fewer than 1,000 feet away from its base. But, as it happened, the South Tower fell first, crushing itself floor by floor, spewing a cataract of rubble out in all directions. Johansen felt the ground shake beneath his feet and within seconds the waterfront was buried beneath half a foot of burning, itchy powder.

The ground did indeed shake. The force of the collapse, pulsing into the bedrock and rippling out to the Atlantic Ocean and along the Hudson's riverbed, registered on instruments at a Columbia University Observatory, more than 18 miles from New York City, as a magnitude 2 earthquake.

Helmsman Carlos Perez watched it happen from the water. From aboard the Coast Guard utility boat that he'd been patrolling around the tip of Manhattan, sweeping back and forth from North Cove south around the Battery, he heard thunder echo across the harbor. Perez and his crew looked up to see 2 World Trade Center collapse under its own weight, releasing a torrent that poured like magma through the empty spaces between buildings before spilling out over the water. "The rumbling became so loud that I couldn't hear myself yelling at my crew to hang on as we tried to outrun the debris cloud that engulfed us," Perez recalled.

His mind flashed to a radio transmission from Activities New York that he'd heard a short time earlier issuing a stark warning.

Due to the high concentration of heat, building engineers had cautioned that the towers might collapse into the Hudson River creating a 40- to 50-foot wave. Steering the boat to what he hoped was a safe distance away, just south of Governors Island, Perez awaited instructions. But the radio had gone silent. Manhattan Island disappeared behind a veil of white.

Each subsequent event amplified the crisis unfolding at the World Trade Center, intensifying the fear and panic and increasing the numbers of people directly caught up in the catastrophe. With the avalanche of toxic dust and debris came terror. Bridges and tunnels were closed. Streets were clogged with stalled traffic. No trains were moving. Suddenly, hundreds of thousands of visitors, residents, and commuters found themselves trapped in Lower Manhattan, struggling to grasp what was happening and trying to answer one question: How could they get off the island?

PART TWO

THE EVACUATION

Courage is a kind of salvation.
—Plato

Chapter 4

"I was gonna swim to Jersey."

THE LONGSTANDING TRADITION of mariners assisting those in peril is as ancient as seafaring itself. Stemming from a moral duty rooted in pragmatism about the implicit dangers of nautical life, the obligation was signed into U.S. admiralty law in the aftermath of the April 15, 1912, sinking of the RMS *Titanic*. Codified as 46 U.S.C. section 2304, the law provides: "A master or individual in charge of a vessel shall render assistance to any individual found at sea in danger of being lost, so far as the master or individual in charge can do so without serious danger to the master's or individual's vessel or individuals on board." Failure to comply is grounds for criminal sanctions.

This rule, however, did not apply to the situation in Manhattan on September 11. At least not technically speaking in most cases. But that didn't stop the boatmen and boatwomen from New York harbor and beyond from feeling compelled. At stake were notions of identity, of mariners' acclimatization to taking and mitigating risks, of what disaster researchers James Kendra and Tricia Wachtendorf call "professional honor." Mariners they interviewed in the aftermath of the attacks did not talk of *choosing* to help. Instead, without planning or protocols, many undertook the evacuation out of a sense of duty, unquestioningly, applying to this land-based calamity their mandate from the laws of the sea. The compulsion to rescue, stitched into the fabric of nautical tradition, propelled mariners into action, as did the sense, for many, of New York harbor—waterfront included—as "home."

Staten Island Ferry Captain James Parese was sprinting across the upper deck of *Samuel I. Newhouse* when all at once the air turned "a very weird color, like a greenish gray." He'd been preparing to pull the orange ferryboat away from the slip at the southern tip of Manhattan when a blanket of gypsum dust, smoke, and ash blotted out the sun. The 6,000-passenger ferryboat, one of the highest passenger capacity vessels in the world, was filled with people desperate to evacuate the island—some panicked and crying, some bleeding, some with no shoes—and now they scrambled for life preservers, thinking the boat was on fire. It was one minute before 10 A.M. The South Tower had just collapsed.

Parese's eyes and throat started to burn. "I remember looking out towards Jersey and I couldn't see anything." For a moment he questioned the decision he'd made earlier that morning to set out on this rescue mission from the safety of Staten Island. But when pieces of white plastic began drifting down from the sky, it reminded him of snow. He felt suddenly serene.

"You know when you're a kid and you're walking in that gentle snow and it's very quiet and peaceful? That's kind of what it brought me back to. . . . I was completely calm at that point. . . . All I could do was focus." As the captain steered the 300-foot, 3,335-ton ferry into a harbor crowded with other vessels—navigating by radar with zero visibility—the lives of thousands of distraught passengers depended on that focus. Parese drew upon decades of experience as a mariner, a profession where the notion that panic leads to peril is as deeply ingrained as the tradition of helping those in need. He delivered his passengers to safety on Staten Island, then returned, time and time again, to Manhattan to pick up still more.

The shushing noise reminded marine engineer Gulmar Parga of crystal shattering. "Like a giant chandelier . . . It was all the glass breaking." Together, the twin towers contained 43,600 windows and more than 600,000 square feet of glass. The sound of its fracture reached Parga as he stood on the deck of fireboat *John D. McKean,* about a thousand feet away. A split-second earlier, Parga had heard what sounded like a series of explosions. "The floors were collapsing: Boom! Boom! Boom! Boom!" From where the boat was tied up along the seawall just south of North Cove, it looked to Parga like a string of dynamite had been rigged to blow up each level of 2 World Trade Center in rapid succession.

A career firefighter, Parga still grimaces at the recollection of his first thought when he grasped that the tower was falling. "The feeling of guilt was instant, but I felt like, *I'm glad I'm not under it*. You know what I mean? I was out there on the water and it was collapsing and I knew all my friends were dying in there—possibly Tom Sullivan and possibly the captain—they were dying and I thought, *I'm glad I'm not under it*."

He watched a colossal dust cloud rise up from where the tower once stood. "Like a nuclear explosion, it's spiraling and spiraling and coming closer and closer to us."

Scientists have since calculated the energy released in the collapse as equal to about 1 percent of a nuclear bomb. Because energy cannot be destroyed, only transformed, the work that construction crews had put into building the 1,350-foot tower (lifting sections of wall, pouring concrete, bolting steel) had been stockpiled in the South Tower for three decades. When the 500,000-ton mass—all that steel, all that concrete, the contents of nearly 5 million square feet of office space, including people—came down, it released 278 megawatt hours of potential energy—enough power to supply all the homes in Atlanta or Oakland for an hour.

Seconds later, the wall of dust and debris descended upon the boat. Parga dropped to his hands and knees. "Whoosh, it went from daytime to nighttime. It got dark instantly and you couldn't breathe. I could only describe it like sticking your head in a bag of sawdust and shaking it and trying to breathe. You imagine all that grit and everything going in?" Parga waited to die. Instead of sawdust, Parga, like so many hundreds of thousands of people caught in the cloud, was actually breathing in asbestos, lead, glass, heavy metals, concrete, poisonous gases, oil, and exploded jet fuel, as well as the pulverized contents of hundreds of offices, including humans. A few choking breaths later, Parga realized the cloud had not killed him. He found he could actually manage to take in some air.

Gritty, gray sediment mantled every surface in the plaza around North Cove, from the tree leaves, bench slats, and paving stones to the lampposts and railings at the seawall. Thousands upon thousands of bits of paper tumbled softly through the drifts—documents depicting daily stock trades alongside mementos of individuals' most meaningful moments. One sheet displayed a blur of crunched numbers and corporate acronyms. Another held a newborn's tiny inked footprints. But amid the maelstrom of fear that accompanied the onslaught of debris, these details went largely unnoticed.

The plaza had devolved into utter panic. For many people caught up in the cloud, the impulse to flee overshadowed rational thought. And when they invited peril by jumping, onto the boat or into the water, the crew of fireboat *John D. McKean* did their utmost to try to help them.

Technically speaking, the Hudson is not a river, but a tidal estuary wherein fresh mountain water from its source—Lake Tear of the Clouds, sitting atop Mount Marcy, which is New York State's highest peak—mixes with salt water from the sea. On its journey to New York harbor, the Hudson travels 315 miles and drops 4,322 feet in elevation. Long before it was named after explorer Henry Hudson, the river was called by its Algonquian name *Muhheahkunnuk*, meaning great waters constantly in motion, either ebbing or flowing. For half of each day, the Hudson acts like a river, flowing downstream. But every six hours the water switches direction as Atlantic seawater rushes upstream, raising the water level an average of 4.6 feet during the flood tide before the ebb tide drains it back down. Though few people except mariners usually give them much thought, the rise and fall of the tides played a key role in the fates of people who ran until they'd run out of land.

Low water at the Battery had come at 8:50 that morning. The outflow of the ebb tide had left a drop between the seawall surrounding North Cove and the river's surface of five to seven feet. The steep swoop of fireboat *McKean*'s bow lined it up almost flush with the top of the wall, but left a farther drop at the stern. When the dust cloud descended, people spilled over the wall onto the boat. As quickly as they could, the crew scrambled to set up the three ladders they had aboard to receive them. But the ladders weren't enough.

"We're starboard-side-to, facing north, and I'm helping everybody I can within my scope," recalled marine engineer Gulmar Parga. "They're climbing over the railing. They're handing everything over the railing down to us. . . . All the people are jumping on the boat on the bow, and they're jumping on the stern. And I can't stop them."

One middle-aged woman stood outside the railing atop the seawall, calling down to the crew. "Help me. Help me!" she cried as she lined herself up to leap onto the deck.

"Wait. Don't jump. Wait!" urged the firefighters as they quickly assembled to try to break her fall. "So we got two guys," said Parga. "'Okay, jump!' And she'd land on top of you, knock you down. And then we'd get up and help the next person: 'One, two, three, jump!'"

When they spotted a man in a wheelchair at the railing, several firefighters worked to carry him and his chair aboard. Another man stood above the stern, poised to jump. "Don't jump," yelled Parga. "Wait! Don't jump!" But he didn't wait. When his foot hit the steel deck, his leg snapped at the shin. He howled, but Parga couldn't stop to attend to him while people kept pouring over the side. "He had to stay there and scream."

Some panicked evacuees began untying the mooring lines until the crew stopped them. "We can't untie the boat. We've got more people," urged pilot Jim Campanelli. Then he continued hollering reassurances to the people atop the seawall. "We'll get you out of here," he said. "We're not gonna leave anybody back." Within minutes nearly 200 people had boarded, a number of them sustaining severe injuries.

One minute before ten o'clock, the dinner-boat captain, Jerry Grandinetti, had just stepped inside the VIP Yacht Cruises office when he heard what sounded like a third explosion. He slammed the door behind him. "The world went black," he explained. "And I mean totally black." He and another guy in the office quickly set about shutting off the air conditioners to make sure that whatever was outside didn't get sucked in.

A few moments later, when the blackness faded to a hazy gray, Grandinetti went back out and started walking east toward the trade center. The whole area was blanketed with half a foot of ashy powder and littered with metal and other debris. Grandinetti could see 20 to 30 feet in front of him, but he couldn't see up at all.

So while millions of people around the world had watched the South Tower collapse, Grandinetti had no inkling that it had fallen. The same was true for countless others who were closest to the unfolding disaster. Even people still in Tower One, both rescuers and civilians, didn't know the other building was down. The very idea was beyond conception, and little reliable information penetrated from the outside world.

At the foot of Gateway Plaza, Grandinetti watched a handful of dazed people wander west toward the water. Then two firefighters emerged from the fog carrying a third who was unconscious. Grandinetti guided them to the office where they laid the man on a table. Beyond that they exchanged few words. Shocked silence and stoicism steered their interaction.

Grandinetti did the only thing he could think of. He grabbed a case of water from the supply closet to hand to people on the street and headed back to check on the boats once more. As he reached the north side of North Cove, two other firefighters intercepted him. "Can you get that dinghy running?" they asked, referring to the yacht company's 20-foot inflatable Zodiac. They asked if he could run them south near the tip of Manhattan so they could meet up with their fire company. "Let's go," Grandinetti replied.

The dinner-boat captain insists he hadn't been "looking for stuff to do" to help in that moment. But once asked he agreed with no second thought. "I just knew I had the ability." So he welcomed the firefighters aboard the dust-covered, "center-console, big-engine, way-too-fast" rigid inflatable and headed south. Nearing the Battery, Grandinetti popped the boat into reverse to slow his approach into the slip at an unused ferry terminal. Then the engine stalled.

There was no mistaking the heart-sinking lurch that had stopped the boat dead in the water. When the Zodiac quit, just feet off the dock where Jerry Grandinetti had intended to drop his passengers, the seasoned boatman recognized instantly what had happened. The braided polypropylene bowline had wrapped itself around the propeller.

By this hour, just after 10 in the morning, the harbor was filled with ferries, NYPD harbor launches, and other vessels running around at top speeds. All these boats kicked up even more wake action than usual and now, dead in the water, the rigid inflatable bobbed like a bath toy in the three-foot chop. Waves or no waves, Grandinetti had no choice but to hang himself over the stern to try to untangle the rope from the propeller blades. Each time the Zodiac dipped between swells, he got dunked underwater face-first. Down. Up. Down again. Grandinetti was dipped repeatedly like a sugar cube in salty Hudson River tea, as he worked to free the propeller. And still the prop stayed jammed. Finally, line removed, he reached the dock, the firefighters debarked, and Grandinetti pulled out, rounded up, and headed north.

On his trip down, the wind had been at his stern. Now that he was northbound the dust covering the boat kicked up with each new wave, plastering his wet torso, face, and hair. Though he was wearing and breathing in the pulverized remains of a 110-story tower and its occupants, Grandinetti still didn't know that 2 World Trade Center had fallen.

Then Grandinetti spotted his friend, fellow captain Pat Harris, attempting to pull the sailing yacht *Ventura* up to the seawall just north of Pier A in Robert F. Wagner Jr. Park, where hundreds of people pressed up against the railing at the river's edge. When the first tower fell, Harris had been safely tied up a mile away in Morris Canal, across the river in New Jersey, attempting to get fuel. But when someone on land announced that people were jumping off the seawall in Manhattan, Harris had cast off lines and set out once again, passengers in tow, to offer what aid he could.

He hadn't come across anyone in the water, but found swarms of people desperate to get off the island. When he tried to maneuver his yacht into position to board passengers he discovered the river was too choppy. "We got smacked by some wave action and I realized that the *Ventura* was just the wrong boat for that." Although Harris conceded that "going from a wall to a bowsprit [the spar extending forward from the ship's bow] on a moving, bobbing boat" was "not a good boarding process," one man (a 30-something St. Louis resident who worked for the news agency Reuters, the captain recalled) did manage to climb aboard. When he saw Grandinetti, Harris called him over to ask if he could use the little Zodiac to ferry passengers from the seawall over to the *Ventura* since boarding them directly was sketchy at best.

"It's too rough, Pat," Grandinetti replied. "It's too rough." Plus people would have had to climb over the metal railing and then jump down well over five feet into the boat.

"I remember Jerry saying, 'These people are in no danger here,'" Harris recalled. "And it made a lot of sense." Harris told the would-be evacuees that other boats better suited to boarding were on the way, then headed for New Jersey to deliver his existing passengers—the mate and his family—out of harm's way.

Volunteer firefighter Bob Nussberger thought he must be dead. The sudden silence, the absence of all sensation, had left him with an odd sense of peace. Where the sun had shone in a cloudless blue sky there was now just blackness. In the space once filled by shrieking sirens only dead air remained. All at once there was nothing. Minutes ticked by.

Just before 10 A.M., Nussberger had trudged in full firefighter's bunker gear through the gray soot and papers strewn across

West Street. He and his volunteer fire company had been called to the World Trade Center from Broad Channel, an island in Queens, near John F. Kennedy Airport. Immediately upon arriving, Nussberger had received orders to tend to burn victims in a South Tower elevator. To find out which elevator, they'd been told to report to a lieutenant in the lobby of the Marriott Hotel in 3 World Trade Center, the 22-story building tucked between the two towers.

As they walked across West Street, traces of the carnage were unavoidable. Amid the burning rubble, shoes, airplane parts, garments, and luggage strewn about in the plaza lay charred body parts. But to Nussberger, the remains didn't look real. It was all too much to fathom. "But still and all," he said, struggling to find the words, "Just the idea . . . You knew."

Before he and his unit could step beneath the slanted glass awning at the hotel entrance, a police officer halted them with an outstretched palm and a grim expression. He'd been standing there, working the door, watching overhead. "Okay," he'd say to people trying to flee the towers through the Marriott's Tall Ships Bar and Grill. A few people would sprint across the street. "Okay. Now," he'd yell, and a couple more would dash across. As he gestured to Nussberger, stopping him from entering, the officer pointed up. "There was a lot of debris coming down," Nussberger explained. "Use your imagination on debris."

Though Nussberger didn't know it, a half-hour earlier firefighter Danny Suhr had become the first FDNY fatality of the day when he was struck and killed by a falling body as he approached the base of the South Tower. FDNY Captain Paul Conlon, of Engine Company 216, had been walking ahead of him at the time. "It was as if he exploded," the captain recalled in an FDNY interview conducted four months later. "It wasn't like you heard something falling and you could jump out of the way."

Standing a few yards away from the hotel lobby entrance, the Broad Channel volunteers had waited for the sky to clear, then managed to proceed five or six feet before a roar erupted above them. "By the time we heard that building coming down, it was already halfway down," said Nussberger. "I didn't get the chance to run." The pancaking of the tower, beginning at 9:58:59 A.M., had taken all of 10 seconds.

"I don't know how high I was off the ground, but I knew I wasn't on the ground," Nussberger recalled. Then, with "a large thump," he hit something solid. "At that point there was no more noise. There was nothing. There was a dead silence. I kept thinking to myself, *If I'm dead, that was a nice peaceful way of going, you know?*"

Soon the blackness faded to brown. *I'm outside*, he thought. When he tried to move, a hot rush of pain seared through his shoulders, his chest, his face, his feet. The throbbing and stabbing told the 59-year-old he must not be dead, so he tried to right himself. Whatever was piled on top of him was thick but not too heavy to push aside. He crawled across hunks of blasted concrete and lengths of twisted steel, sucking in air so thick it was like trying to breathe sand.

"There was a lot of junk in my mouth," said Nussberger. "My eyes were on fire. There was blood all over my hands and my face." Moments earlier he had seen 80 or 90 people rushing around the city streets. Now 20 or so soot-covered zombies emerged from the wreckage. A blizzard of paper continued to tumble through the sky. "I kept on telling the people: 'Get into the building! Get into the building. It's unsafe. There's still a lot of stuff coming down.'"

Lights flashed and sirens blared all around him as FDNY firefighter Tom Sullivan headed inland on the hunt for fireboat *John D. McKean*'s captain. Near the southwest corner of Liberty and West Streets, directly across the street from the Marriott, an engine company was working to extinguish a street-level blaze—a car fire, maybe, or a large rubbish bin aflame. Burning wreckage littered the street.

To Sullivan's right, a scatter of FDNY equipment—safety-red pumpers and aerial ladders—and an army of white ambulances stood parked beneath the enclosed pedestrian walkway crossing West Street. The diversity of their detailing revealed the wide reach of the emergency medical services (EMS) response. Among the rigs was the ambulance that Nussberger and the Broad Channel volunteers had driven into Manhattan.

Up ahead, about a block to the north, Sullivan spotted several "white hats," or chiefs, at a command post directly across from the North Tower. Among the top brass standing there was 33-year FDNY veteran and Chief of Department Peter Ganci. He had just received a message from a Buildings Department engineer warning that severe structural damage to the towers meant they might be in imminent danger of collapse. Then an awful rumble shook the earth.

"What the fuck is that?" asked Ganci, as the upper floors of 2 World Trade Center began to cave.

Standing directly across from the disintegrating South Tower, Sullivan looked up to see a wall of gray dust barreling down on him. He did a quick about-face and ran, but didn't get more than 15 or 20 feet before a force that "felt like a giant hand" pushed him from behind, slamming him onto the pavement. His helmet went flying. The fall shredded the skin on his palms and elbows. "I was lying facedown in the street now in total darkness," he recalled. A hail of broken-up concrete teemed down from the sky, pelting Sullivan on his back and his bare head as he blindly tried to reorient himself. The stinging powder burned his eyes and plugged up his ears, his mouth.

He groped around for his helmet but quickly gave up. He managed to lay hands on his flashlight, but its beam barely penetrated the soot- and gypsum-clogged air. So he focused on feeling for street curbs, knowing that heading due west would lead him back to the water and to the fireboat. Scrabbling beneath an ongoing shower of rubble, Sullivan at one point resigned himself to the futility of his efforts. *There's no way of getting out of this*, he thought, picturing the building toppling over instead of pancaking down, reasoning that heavy steel would soon fall, crushing him where he crawled.

But he kept inching forward, pulling his shirt up over his nose so he could breathe. "I got to the next intersection and just kind of lined myself up to go straight across the street and catch the next corner." On the other side, Sullivan felt the avalanche lessen. He bumped blindly into parked cars. He could hear people talking, hollering, screaming.

Still on his hands and knees, he found himself cornered, momentarily, in a notch at the base of the Gateway Plaza building. Then he spotted a flashlight. He grabbed at it but quickly let go when he realized he was gripping the muzzle of a rifle. The man holding the gun identified himself as an NYPD ESU officer. Sullivan could feel his helmet and vest. The police officer, too, was looking for a way out.

"Listen, follow me," said Sullivan. In time, a dim light penetrated the haze and the firefighter was able to recognize the ramp that led down to the waterfront plaza as the same one he'd climbed moments earlier on his way inland.

By the time he reached the Hudson, the air had cleared a bit. The first thing he saw was the *John D. McKean* nameplate on the bow. *Thank God they didn't leave me.* Already the fireboat was overrun with people, civilians desperate to escape the cloud. Sullivan climbed aboard and headed for the nearest sink to flush out his eyes, which were gummed up with particles. Then he returned to the deck to help the crew manage the flood of people hurtling onto the boat.

A group of women—mothers or nannies—handed down the infants in their care before climbing down themselves. Sullivan carried several babies down to the crew's quarters below decks, lining them up in a bunk "like little peanuts."

An older woman seemed to have landed headfirst on the deck, cracking her skull. Blood trickled out of her ears and nose, pooling where she lay by the hose reel on the bow. It was impossible to know the extent of her injuries, but it was clear that she needed medical attention—and the only way to get it was to cross the river.

"*I was gonna swim to Jersey*," recalled telecommunications specialist Rich Varela. "That's what I was thinking." He had just watched people falling through the sky and now he tried to shake off the images as he headed west down Liberty Street from

1 World Financial Center toward the Hudson. His cell phone rang. A friend announced that another plane had just hit the Pentagon. *This is completely out of control.*

Halfway to the river, still reeling, Varela heard a rumbling. He turned around to see the South Tower implode and started to run. He didn't stop until he had his hands on the steel railing separating the water from the land. But before vaulting over the rail he turned around to look. *Which direction is the building falling?* He knew there would be no escaping if the tower toppled toward him like a pine tree. Instead he saw that it was collapsing within its own footprint.

As the plume of ash, dust, and smoke barreled toward him, Varela noticed a boat at the seawall. He bounded over the railing and leapt onto the bow of fireboat *John D. McKean*. When he hit the deck, Varela felt his leg buckle and almost snap. A mass of people jumped on after him, falling onto the deck, landing on him, and the boat rocked under the weight of the leaping hoards. Varela worried it might capsize. He stumbled over people on his way to the far side of the deck, away from the avalanche, then curled in on himself, choking as everything went black.

When the air cleared a bit, Varela saw casualties all around him. Somebody was nursing a broken leg. Others complained about pain in their arms. He saw the same woman that Sullivan had noticed, splayed out beside the bow pipe. It looked as if she had landed face-first on the steel deck. He hollered to a nearby firefighter that she needed medical attention—that she was unconscious and might already be dead. There was little anyone could do on the boat, so reaching a triage center, quickly, was imperative. But other lives needed saving, too.

Still coughing and gagging, Varela yanked off his gray-green long-sleeved cotton shirt, tore off a strip, and wet it with water he found dribbling from some leaky hose on deck. He tied the makeshift filter around his face and then tore off more strips for a few fellow passengers. He heard commotion near the bow on the starboard side. People were yelling. Someone was in the water.

Tammy Wiggs had kept her eyes fixed on the burning buildings while she trailed Karen Lacey, walking west from the New York Stock Exchange. "The whole sky was littered with things falling." At first she thought everything fluttering through the sky was only paper. Then she spotted movement and recognized the shapes as people. "I saw dead bodies. I saw bodies jumping out of the building."

Like "a little puppy dog" Wiggs continued following Lacey. "I thought she must know what she was doing because she was so certain about getting home. . . . I didn't realize it meant going that close to the buildings." She continued to dial her sister's number. "I wanted her to be the person that was taking care of me at that point and not somebody that I barely knew," she explained years later, through tears. "I just wanted her to tell me what to do." What she couldn't know then was that the corner of 3 World Financial Center where her sister's Lehman Brothers office was located would soon be sliced by fragments of a crumbling trade center tower.

At one minute before ten o'clock, Wiggs and Lacey exited the winding paths of Battery Park City and stepped onto the plaza at the southwestern corner of North Cove. Their destination—New York Waterway's World Financial Center ferry terminal—now lay just 500 feet away. They faced 2 World Trade Center head on. Wiggs was looking directly at the South Tower when it began to fall.

She didn't hear a noise or feel the ground shake. Instead she noticed the building suddenly growing smaller. Seconds later the pulverized tower erupted out onto the plaza. Desperate for cover of any kind, Wiggs and Lacey ducked behind a ticket booth—a structure so small it reminded Wiggs of the old-fashioned phone booths she had seen in London during her study abroad. "I looked up and saw the mass of what was coming down," she said. "I realized, this is not going to do anything to help." So she shot for the water's edge.

While Wiggs may have been a novice at her job, she was no novice around water. In fact, she had been a lifelong sailor. "If anything, the water's calming to me," she said, explaining that she had sailed seashells and blue jays as a child, then advanced to 420 sailboats, which she'd raced in the Junior Olympics and in the Women's World competition in Yokohama, Japan. "Growing up a sailor, you live a sailor." She recalled beginning mornings by listening for flags flapping—the sound that would tell her whether she'd spend the day bobbing aimlessly through a windless sky or whipped around so fast she'd flip her boat and wind up swimming.

The breeze on this warm September morning was "probably blowing at least 10 knots," recalled Wiggs. "If it hadn't been a windy day then I wouldn't be here today."

When a piece of burning rubble flew at her face, Wiggs fled to the river by instinct. "It was the opposite of where the whole mess was coming from," and water seemed like the only way out.

She and Lacey clambered over the three-foot-high steel railing atop the seawall. A police officer standing on a nearby stone light pillar cautioned them: "Ladies, don't jump." Wiggs's first impulse was to obey. After all, he was a police officer. Maybe jumping into New York harbor wasn't such a good idea.

But by the time the two women had made it over the railing—its top section curved inward, toward the sidewalk, making it a particularly awkward climb—the officer amended his position. "Don't jump . . . yet," he said.

Perched on a seven-inch-wide ledge, suspended some five to ten feet above the river's surface, clutching the railing behind her back, Wiggs pondered his last word: "yet." That "yet" constituted equivocation. It revealed that no one, not even a police officer, had any idea what was coming next.

The plume of gray-black smoke had thickened and swelled, mushrooming up as the building crumbled down. The wave of dust cauliflowered up and out, shattering windows and blowing out the immense panes of glass in the dome of the nearby Winter Garden. The mass barreled down, and Wiggs struggled to draw breath.

She gulped what air she could from the choking cloud that had engulfed her, then released her grip on the rail at the small of her back. She reached blindly for the hand of the woman beside her.

Then, together, they jumped. Time stretched as the two plunged through the dense black air. Split seconds sprawled into an eternity.

A fraction of an instant later, Wiggs's five-foot-three-inch, 120-pound frame plunged into the river. Cold seized her chest. Frantic to flee, she kicked back to the surface, stretching her body long, clawing at the water. But after a few strokes she faltered. The current was ripping. It tugged her downstream. The pall of black smoke thick with particulates—a clenched hand covering her nose and mouth—left her gasping. The airless sky above the river's surface offered no oxygen, so Wiggs dove. "I couldn't breathe at all because of the smoke and particles. I went under the water hoping to find oxygen," she explained. "It doesn't make any sense, I know, but that was the desperation: *There's no oxygen left in this air so the only other thing to try is under the water.*" She kicked off her shoes, but clung to her purse. "I knew my ID was in there. I wanted my parents to have my ID to identify my body."

The drop was farther than Wiggs had expected and the water cold and salty. "I thought we were jumping to swim across," she explained years later, conceding that the very idea seemed embarrassingly absurd to her now. "Running wasn't going to be fast enough and for some reason I thought swimming was going to be faster." She also thought that being in the water could somehow protect her from the cloud. But when she kicked back to the surface, the air felt thick, heavy, solid. She bobbed up and down three or four times on her fruitless search for oxygen underwater. Then, just as she thought she "didn't have another breath in me" the sky began to clear.

Her original impulse to swim the mile to New Jersey had brought her about 10 feet from the Manhattan shoreline. Now she realized how quickly the current was dragging her out to sea, and that she couldn't breathe well enough to swim.

Even strong, experienced swimmers who've *planned* to take a dip into the Hudson's waters have been known to misjudge the challenges. The first person to swim around Manhattan Island was 18-year-old Robert Dowling. On September 5, 1915, his circumnavigation took 13 hours, 45 minutes. Today, synchronizing their routes to maximize the tides, swimmers can circumnavigate the island in half that time. Races around Manhattan Island, across the English Channel, and between Catalina Island and the Southern California coast are considered to be the Triple Crown of open-water swimming—demanding tests of physical and mental stamina.

But even much shorter swim events, such as Cove to Cove—the annual half-mile swim from South Cove to North Cove in Battery Park City—are timed to make sure swimmers don't have to fight the powerful currents. One experienced swimmer who participated in the 2004 Cove to Cove explained that she'd started the race feeling confident, encouraged by reports of the evening's strong favorable current, but soon realized she had "completely underestimated" her Hudson swim, explaining that any benefit she might have gained from the current was counteracted by the choppy water. The river's vastness had left her floundering and disoriented (accidentally swimming east toward the seawall instead of north) and having difficulty gauging how far she'd actually swum from one moment to the next. Others have likened swimming in the Hudson to enduring a washing machine's spin cycle.

And now, here was Wiggs, in over her head in more ways than one. When, finally, the air lightened from coal-black to haze gray, Wiggs spotted the pointed bow of a boat, its maroon hull about 15 feet away. Thinking that the vessel was under way, she panicked. She worried about getting chopped up by the propellers. "We're in the water! We're in the water!" she screamed, over and over.

"Who's in the water?" a voice called back through the fog.

Then Wiggs noticed a thick rope stretched overhead. It was the bowline leading to the boat where the man stood, calling down. The boat was not under power after all. With a few strokes, she made her way back to the seawall, bloodying her hands as she clawed at the rough cement surface, struggling to secure a grip. Finally her fingertips located a beveled-edge joint where two slabs met. She dug her close-clipped nails into the notch and kicked to tread water, battling the pull of the current.

CHAPTER 5

"It was like breathing dirt."

TO PROTECT HIMSELF FROM THE DEBRIS STILL FALLING after the South Tower's collapse, volunteer firefighter Nussberger climbed through the entrance of a domed office building with blown-out windows and a green marble lobby strewn with ashy wreckage. Once sheltered, he struggled to piece together the details. That thump he'd heard in the split second before the world went silent and still, when the air had suddenly become too thick, too solid to transmit the vibrations that humans perceive as sound—just before the wave of agony that later roused him—was the impact of his body being flung against the side of this building, 2 World Financial Center, 70 feet away from where he'd last stood.

At some point Nussberger made his way out of the rubble-filled building and wandered back outside. He began stumbling across the waterfront plaza, struggling to breathe. *I'm gonna have a heart attack. I know I'm gonna have a heart attack.* He was fading in and out of coherence—crawling, walking, and falling again as he crossed an outdoor plaza until eventually he found his way down a ramp to the Hudson River. He collapsed again on the hexagonal paving stones along the water's edge. This time he couldn't get up. The adrenaline that had carried him thus far seemed to have quit.

Against the seawall in front of him, a fireboat loaded with evacuees was preparing to pull out. Some men gathered on the shoreline spotted Nussberger collapsed in the dust in his turnout gear. They heaved him over the railing and onto the boat below, one gripping his arm so hard that it left a full handprint bruised into his skin. Two men helped him to the bow where he lay down atop a pile of fire hose. Disoriented by pain, Nussberger looked up at the huge bow pipe above him and tried to piece together where on earth he could have ended up.

A short time earlier Richard Larrabee, commerce director for the Port Authority of New York and New Jersey, had wondered the same thing. He had been standing in the lobby of the Marriott Hotel. Now he was crawling through wreckage on his hands and knees. When light finally penetrated the darkness he looked up and saw a street lamp. *Where the hell is there a traffic light inside the World Trade Center?*

Then it dawned on him that he was outside. He had wound up on West Street. Somehow, he had been spared when the hotel that sat sandwiched between the two trade center towers was cleaved from top to bottom "as if a giant scissors had snipped the building in two," as reporters would later describe it. While reinforced beams installed following the 1993 bombing had protected one half of the lobby, the other half was crushed, taking whole fire companies—at least 40 firefighters—along with it.

All morning the Marriott Hotel had served as a critical escape route allowing close to a thousand people fleeing 1 World Trade Center to leave without facing the peril of walking outside. Since the first minutes of the disaster, hotel staffers had been ushering people to safety. They'd also helped redirect firefighters and police officers who, unfamiliar with the layout of the trade center complex, had chosen the most visible entrance.

By the time Larrabee arrived, just before ten o'clock, the flow of evacuees had slowed. The lobby was a blur of black turnout gear as the sea of helmeted firefighters who'd assembled awaited orders. Sirens wailed. The apprehension was palpable. Still, the hotel had seemed a safe enough distance away from the fires for Larrabee to gather his senior staff so they could make critical decisions about next steps.

After a 34-year career in the Coast Guard, from which he had retired as Boston's district commander, Larrabee now had a civilian job that carried enormous responsibility. Every commercial transaction in the Port of New York and New Jersey, the largest container port on the East Coast, was under his domain. He knew Port Authority activities would surely suffer disruptions as a result of what was unfolding at the World Trade Center, and the financial stakes were high. Larrabee's job was to anticipate and mitigate problems as best he could. But before his team could even begin to formulate a plan, the world had changed entirely.

Seventy-two minutes earlier, from his office on the sixty-second floor of the North Tower, Larrabee had felt the building shake so hard that the desks shimmied and framed photos fell off the walls. None of the hundred or so Port Authority staffers had hesitated before gathering their things to leave. "The reaction on everyone's part was, 'Let's get the hell out of here.'"

Though Larrabee had only been working in the building for a year and a half, most of his employees had been present in 1993 when power outages and thick smoke conditions from the explosion in the parking garage had left people in the North Tower blindly groping and choking their way down as many as 106 flights of stairs in the dark. Larrabee understood that his staffers "were already sensitized to the fact that things could go wrong in that building, and that they were up at a really high level." Even on a clear day like this one, the 22-inch-wide windows separated by spandrels and columns that architect Minoru Yamasaki had designed to palliate his fear of heights offered good visibility out, but not down. From the sixty-second floor, it was impossible to see the ground, Larrabee explained. "You just physically couldn't see it. You couldn't tell if it was raining or what it was doing outside."

But on this sunny Tuesday morning, the flaming debris that he watched plunge past the windows on the southwest corner of the building pressed the point that it was time to go. Without knowing that a plane had hit the building, Larrabee picked up his briefcase, wallet, and cell phone, "which I didn't really carry very much in those days," and headed for the stairs. The stairway was well lit and only slightly smoky, and as the Port Authority group descended, workers from different floors converged.

People offered help to those who needed it, and there was no panic. As the evacuees stepped down, firefighters trudged up.

In the years since the bombing, the Port Authority had spent more than $90 million on upgrading physical, structural, and technological enhancements to the twin towers as well as improving fire safety plans and reorganizing the fire-safety and security staffs. Building upgrades had included installing a duplicate power source for fire alarms, emergency lighting, and intercoms, while safety plan improvements included conducting evacuation drills every six months. Due in part to these preparedness efforts, 99 percent of the towers' occupants on the floors below the airplane crashes would manage to escape. In the aftermath of the September 11 attacks, a number of people would report that changes to the buildings' infrastructure and systems had, in fact, facilitated their ability to evacuate.

It took 38 minutes for Larrabee to reach the ground level. Conversation in the stairwells informed him that a plane collision was what had caused the tower to shake. "One of the things that you'd heard if you worked in that building was that even if an airplane hit the building it wouldn't fall down," he explained. "That was the thought in my mind: *As long as I get down to the bottom I'm going to be fine.*" Like so many others that day, Larrabee believed the mythology that had been established during the earliest days of the World Trade Center's development.

Establishing the towers' ability to survive a plane crash had been just one of many challenges facing the Port Authority during the trade center's decade-long planning and construction process. The idea for the enormous development project had been inspired by a 1939 New York Worlds Fair exhibit called the World Trade Center, which was dedicated to the concept of world peace through trade. As David Rockefeller, grandson of Standard Oil founder John D. Rockefeller, envisioned it, the complex would become the centerpiece of a revitalized Lower Manhattan. To execute his vision, Rockefeller, in 1960, contacted the Port of New York Authority, an agency chartered in 1921 by

New York and New Jersey to build and operate all transportation terminals and facilities within a 25-mile radius of the Statue of Liberty. The Port Authority soon took on the project.

Early on in its development, the complex was conceived not just as the heart of the financial district, but as a major intermodal transportation hub for Lower Manhattan. Plans would include integration and enhancement of the existing railway tunnels (essentially cast-iron tubes) running along the soft riverbed of the Hudson. Since 1909, the Hudson and Manhattan (H&M) Railroad (later known as the H&M Hudson Tubes) had been delivering passengers between Lower Manhattan and Jersey City, transporting a staggering 113,141,729 passengers in its busiest year, 1927.

But ridership soon declined significantly—especially following the openings of the Holland Tunnel, George Washington Bridge, and Lincoln Tunnel, all three of which provided other commuting options. In 1927, the Holland Tunnel, connecting Canal Street in Manhattan with Jersey City's Twelfth and Fourteenth Streets, became the first Hudson River crossing for vehicles. The George Washington Bridge and Lincoln Tunnel soon followed in 1931 and 1937, respectively, increasing traffic flow of automobiles and drawing people away from the railroad.

The World Trade Center project encouraged the Port Authority to take over and renovate the struggling H&M Hudson Tubes in return for the rights to build the new massive complex on the land then occupied by the railway's Lower Manhattan terminal. And thus, in 1962, the Port Authority Trans-Hudson Corporation, or PATH, train was born. The Port Authority unveiled, in January 1964, an architectural plan for the trade center site featuring the world's tallest buildings. Two years later, construction began. The first tenants moved into the North and South Towers in 1970 and 1971, respectively, and the iconic buildings, soon filled with household names, came to represent the powerful and elite.

The twin towers became symbols of American achievement, global trade, the ideals of capitalism, and the quest to be the biggest and the best. Thirty-five years later, though the superskyscrapers seemed to have accepted the brutal assault of airliner intrusion, their deconstruction was already well under way.

Only once Larrabee had reached the North Tower's lobby, where firefighters and evacuees sloshed through several inches of water left by the sprinklers, did he hear that the twin towers had suffered a two-plane attack. Not long after he called his team over to the Marriott's lobby for an emergency meeting, Larrabee heard a large explosion, and he was thrown to the floor. "Everything crumbled around us." He found himself lying in the dark, pinned under a heavy weight. When he felt movement he realized the weight was not only rubble but also a person. A man, a firefighter, had landed on top of him. Although they couldn't see each other through the thick, choking soot, they managed to stay together, working to crawl away from where they'd landed. "I wasn't aware of anyone else around me." Eventually the two men worked their way to a place that seemed flat. Then called by a sense of duty that steered them in opposite directions, they "kind of shook hands and parted ways."

Just like VIP Captain Jerry Grandinetti, Larrabee was wearing, and tasting, and choking on its remains, but still he was unaware that the tower had fallen. In a state of shock, he shambled off, clutching his leather Coach briefcase—a gift he'd received in honor of his Coast Guard retirement. "One of the handles had broken off and I was dragging this thing along for whatever reason," he recalled, still puzzled by his own behavior. "I don't even remember what was in it." As he shuffled through the ashy streets, he remembered the Coast Guard office located due south, at the tip of Manhattan.

He arrived to find a chain-link fence surrounding the facility and headed for a gate. There, Larrabee recognized an officer who had worked for him in Boston a year and a half earlier. "Ed, it's Rick Larrabee. It's Admiral Larrabee," said the retired district commander.

"I'm sorry I don't recognize you," came the response. "You're going to have to show me some ID." Larrabee complied and the officer apologized. "Oh, Admiral. Geez. Come on in." The chagrined officer hadn't been able to make out his admiral's features through the dust.

Sound began trickling through the silence. First the screeching of car alarms. Then voices calling, movement, other people on the ground stumbling. Beginning law student Gina LaPlaca felt someone grabbing at her bare ankles. "You couldn't open your eyes, but it was pitch black anyway. It just felt very heavy." She managed to right herself, to plant her sandaled feet on the street, stretch her palms out in front of her, and stagger through the blackness.

If, in the seconds before 9:58:59 A.M., LaPlaca had been looking at the South Tower instead of at her phone that replied to her every dial attempt with "All circuits are busy," she might have seen the sudden swell of gray smoke bursting from the lowest point of the burning chasm on the east face of the seventy-ninth floor. She might have seen the building's trusses give way, cocking the top 23 floors of the building to the southeast before the lower floors buckled, unleashing cascades of choking debris.

Each face of the towers, framed in structural steel, had been built to withstand hurricane-force winds of 140 miles per hour. Even on an ordinary day the wind load was 30 times greater than the force of a Boeing 767 airliner's impact. It's no wonder the South Tower had stood for so long before the steel spandrel members supporting the cross-braced floors, each with four inches of concrete decking, finally gave way.

And when they did, a rush of energy had hurtled toward LaPlaca, engulfing her. "I felt this gust, this force push me. I lost my feet and flew forward and landed on my face." Head on the pavement, she discovered she couldn't breathe—that the air had mass, substance. "It was like breathing dirt." The blackness total. "At first I felt like there was no one around me—that all of a sudden they had all disappeared—or that I had disappeared."

A few minutes later, though, LaPlaca began to make out the shapes of cars, buildings, the scaffolding over the sidewalk. "I had no idea what I was in or what had happened. I just walked." Then, without warning, someone grabbed her arm. "Don't go that way!" he hollered. His grip prevented her from tumbling down the steps to the Wall Street subway station.

"Thank you," she said. "Thank you so much. Where are we? What is going on?"

He didn't have answers but explained that he and a friend were headed for the South Street Seaport. He urged her to come along. "Keep moving," he said. "I know where we are."

Save for the dust on their clothes the men, both in their thirties, looked like they were "going to work at a bank somewhere," recalled LaPlaca. The trio joined the masses flooding toward Manhattan's eastern shore. Some had cuts and bruises. Some walked through the debris-strewn city streets without shoes. Mouths full of grit, the two men longed for water. A nearby vendor in a silver bagel and donut cart had given away his whole stock of beverages, but LaPlaca pulled a water bottle from her bag. After they drank she used it to flush out her eyes, which had started to swell shut.

Before long they arrived at a commercial fish supplier, M. Slavin & Sons, where workers offered up water hoses, paper towels, and landlines to the gathering crowds. After rinsing herself off, LaPlaca stepped into the office to use the phone. When she couldn't reach her mother at work, she called her grandmother on Long Island and asked her to tell family that she was okay. Then she dialed her parents' home answering machine: "I don't even know if or when you'll hear this because I know you're both at work. But I'm okay. I called Grandma. I told her. I'm trying to figure out what I'm doing." Back outside a woman noticed LaPlaca's torn, wet, and filthy shirt. "She sort of took pity on me," and offered up a coral V-neck pullover that she had in her bag.

LaPlaca longed to go home—even if "home" meant the Gateway Plaza apartment she'd just barely moved into. But with that out of the question, she stuck with the men as they headed for an East Side apartment about two miles away. "The plan, if there was a plan, was to get ourselves up there, walk as far as we could, try to get on transit if it was going up there and regroup." They'd try the phone, watch the news, and attempt to find out what was going on.

About 20 minutes into their trek, one of the guys noticed that a gash along LaPlaca's shin, as well as a cut by her elbow, were bleeding quite a bit. LaPlaca had registered the pain but been too stunned and distracted to grasp that these injuries might require medical attention. At his urging, the three stopped at a community center where someone helped clean and bandage the abrasions. Back outside, the settling dust and distance from the disaster site had allowed daylight to return. "The sun was really irritating my eyes. I was squinting," LaPlaca recalled. "And finally, I couldn't keep my eyes open at all."

"Guys, I'm sorry," she said. "I literally can't keep my eyes open, even one at a time." They took her to the emergency room at nearby Beth Israel Hospital where doctors determined that the corneal abrasions she'd suffered required bandaging her eyes shut. This left LaPlaca blinded in a hospital with strangers in a new city and wondering, *Now what?*

New York Waterway Port Captain Michael McPhillips had been standing in the wheelhouse of a ferryboat halfway across the river, when he noticed the South Tower begin to buckle. He'd already made eight round-trip runs ferrying passengers to New Jersey since the first plane hit, while also manning the radio and fielding questions that came at him from all corners.

At 9:45 A.M., when another captain offered to take the helm of *Frank Sinatra* so that McPhillips could focus on operations, the overwhelmed port captain had gladly given up the wheel. He soon wound up conducting operations in the wheelhouse of a different ferry, *George Washington*. Now, seeing the cloud exploding up and out from the shrinking South Tower, he barked out a warning to a captain who had been lining up for his approach to the World Financial Center terminal: "Get the fuck out of there!"

In an instant, the scale of the disaster had magnified, transforming the evacuation-in-progress into a full-blown rescue effort. The cloud rolling past the seawall blanketed the river's surface and blinded boat captains, forcing them to navigate by radar alone. But sometimes even the radar couldn't penetrate the particle-filled air. "We were covered in dust," said McPhillips. "The radar couldn't see through the dust. . . . We were pulling into the dock blind. We were leaving the dock blind. I don't know how it happened, but it happened."

New York Waterway Captain Rick Thornton had departed the Battery just minutes before ten o'clock and was steering north toward Hoboken. As they crossed the river, the crowd aboard ferryboat *Henry Hudson*—which was at, if not over, capacity—went quiet. Then a passenger announced he'd just heard the Pentagon had been hit. Panic spiraled through the crowd as people shouted into cell phones. With both pilothouse doors open, Thornton could hear his passengers' gasps and mutterings, their efforts to grasp the unfolding disaster. Some minutes later, as the boat plied past North Cove, cruising due west of the World Trade Center, Thornton heard a man's voice rise above the din of frantic chatter: "They're gonna collapse!"

Thornton rolled his eyes. *This guy's gonna cause a panic*, he said to himself. *They're never gonna collapse. They're just going to burn out the last upper floors and they'll rebuild them.* Ten seconds later the South Tower began to fold, casting a hush over the crowd. "Everybody on the boat stopped talking, put their cell phones down, and just stared in awe. It was complete silence on the boat. You don't get people off their cell phones," explained Thornton. "Especially New Yorkers. Nobody screamed. Nobody made any kind of a noise. And it was the eeriest reaction you can imagine." The smoke that erupted "was a nightmare." It barreled north and Thornton watched people in the streets sprinting to try to outrun the cloud.

Passengers pushed to the starboard side to see what was happening, and the ferryboat took on a substantial list. The building "came down so majestically it was almost beautiful," said Thornton. "It was beautiful but also terrible to behold." His first thought was of the firefighters. Though he said he's not religious, he instinctively made a sign of the cross, apprehending that "tens, hundreds, thousands, I didn't know how many people, but their lives had just winked out at that very instant." And then he put both hands back on the wheel.

"As all the people were streaming off the boat, every one of them thanked us profusely, thank you, thank you, thank you. As they walked by, they said, 'What are you guys going to do now?'

'We're going back in there.' At that time the first tower collapsed, it looked pretty much like hell on earth. And the people were like, 'I can't believe you guys are going back in there.' But, you know, that's what we had to do."

From the very first moments after the first plane hit, ferry crews had operated as ad hoc first responders. They didn't stop after the South Tower fell. Injured people found their way to the waterfront and ferryboats continued to serve as floating ambulances. Approximately 200 injured would end up transported aboard New York Waterway ferries by day's end. McPhillips recalled a man who boarded the ferry in a white shirt and dark pants that were "literally melted" to his body. "He was covered with the dust. His clothes were melted to him and it started to boil, the skin. It started to swell." The man—perhaps Kenneth Summers, perhaps another of the fireball's victims—had suffered burns over his whole body, and his skin reminded McPhillips of a wax museum statue that had gotten too hot. The ferry delivered him across the river to a makeshift triage center being laid out by rescue workers along the Jersey waterfront.

Despite the unprecedented scale of this disaster, mariners' "jack of all trades" capabilities proved essential in the aftermath of the attacks. McPhillips had begun the morning as a port captain tracking boat schedules. But soon he was pulling glass shards from a man's forearm and then wrapping it with gauze from the boat's first aid kit to staunch the bleeding. In some sense, mariners were no different from other civilians who simply brought their ordinary skills to an extraordinary situation, breaking an unmanageable set of circumstances into manageable pieces and tackling them one by one.

Given that 85 percent of our nation's critical infrastructure is controlled not by government but by the private sector, the "first" first responders in most catastrophes, say disaster researchers, are most often civilians. Yet even as civilians, the boatmen and boatwomen of New York harbor were particularly well equipped to serve the public in key ways.

For better or worse, boats and ships—especially oceangoing vessels—often operate in relative isolation, existing as self-contained entities. If a fire breaks out while a vessel is under way, having crew members skilled in shipboard fire suppression is critical to the survival of the vessel and all those aboard. Similarly, rapid response to a medical emergency becomes a matter of life or death.

Ever since the Steamboat Act of 1852, federal law has required the licensure of pilots and engineers working aboard certain vessels carrying passengers for hire. Since 1942, the Coast Guard has been responsible for overseeing all functions of maritime safety, including vessel inspections and mariner certification.

Although the specific prerequisites have changed over time, attaining a Coast Guard "ticket" requires merchant mariners to complete training and earn certifications in first aid, CPR, and shipboard firefighting, among other specialized areas. Additionally, Coast Guard license testing includes whole sections on vessel safety requirements established in the Code of Federal Regulations (CFRs): a 200-volume set of rulebooks, totaling thousands of pages, that governs every aspect of maritime work. Among the provisions are specific protocols for abandon ship and man-overboard drills. According to CFR 122.520, "The master [the properly licensed individual having command of the vessel] shall conduct sufficient drills and give sufficient instructions to make sure that all crew members are familiar with their duties during emergencies that necessitate abandoning ship or the recovery of persons who have fallen overboard."

So, every month, year after year, during Coast Guard-required man-overboard drills, New York Waterway deckhands had practiced retrieving an aluminum ladder, hooking it through notches in the bow as they hung it over the side, and climbing down the seven feet from the deck to the waterline to help imaginary swimmers begin their ascent. On this morning, however, the people who'd gone overboard were real, and all that training was put to the test. McPhillips and the crew aboard the ferryboat *George Washington* pulled at least six people from the water between North and South Coves—the same stretch of the Hudson where

Wiggs and Lacey had been fighting strong currents. Other Waterway boats did the same.

Because of the safety protocols followed aboard New York Waterway ferryboats, execution of water rescues was "very simple," McPhillips explained. "It wasn't a big deal." At least not procedurally. The ferry crews were ready, willing, and able to quickly pluck out people who ended up waving their arms for help shortly after they hit the river that McPhillips described as "brutal."

After one water rescue, McPhillips recalled, the deck crew didn't bother to stow the ladder. Instead the boat proceeded straight to the seawall north of the terminal where they saw dock builder Paul Amico and others helping people over the railing, down ladders, and onto the boats.

Paul Amico had been in the wheelhouse of a Waterway ferry approaching the World Financial Center terminal when the South Tower gave way. If he'd heard McPhillips's warning as the tower came down, the ferry's captain didn't heed it. Despite the whiteout conditions, the boat kept coming. Amico watched the captain maneuver in by radar.

On their way across the Hudson, the two had been discussing the morning's conditions. "The only thing I was thinking about was, *We're going to need to get people out*," recalled Amico. "I was asking the captain about volumes, who was out there on the river—what boats were out there, what captains were out there." As the main fabricator of New York Waterway's dockside infrastructure, Amico knew most of them by name.

When they reached Manhattan, Amico stepped onto the blue barge that served as Waterway's loading platform, its glass walls and white circus-tent roof plastered with powder. The power had gone out and passengers were panicking. "We need to get off!" "You need to get me off now!" "Don't wait for anyone else," they urged, choking, their faces caked with dust.

Their urgency might have given Amico cause to reconsider disembarking on Manhattan Island while everyone else clamored to get off. But he remained clear in his objective. "Civilians are comfortable on the land. Captains are comfortable on the water. I'm comfortable with one foot on the boat and one foot on the land. That's where I work. So that's where I needed to be." Amico decided that the best help he could provide was to serve as a bridge. "*You've got the fire and police over there. Where I can probably help the best is right here on the water's edge,*" he recalled thinking. "Let the plumbers do the plumbing."

He asked the captain to spread the word that he had a handheld marine radio. He knew that land-based fire companies and police had no means of communicating with the ferryboats, and he aimed to span that gap, helping to direct civilians to the river—their quickest means of escape.

Now that the terminal was overwhelmed with soot, Waterway director of operations Johansen had begun directing waiting passengers farther north along the seawall where the air was clearer. From the wheelhouse of the *George Washington*, meanwhile, Michael McPhillips instructed captains to pick up people wherever they could do so safely. As shallow-draft bow-loaders, the New York Waterway ferries could pull up and take on passengers virtually anywhere, which proved hugely advantageous as the day unfolded.

Amico was well aware that only two New York Waterway ferry terminals existed in Lower Manhattan: the World Financial Center terminal, fewer than 1,000 feet from the twin towers; and Pier 11, located around the tip of Manhattan on the East River. He took it upon himself to help figure out more viable options.

"I'm looking at the water's edge and saying, where else can I get Waterway's boats in and out? That's my concern." Just south of the barge, he ran into a police sergeant trying to calm people and keep them moving north. Amico explained that he had communication with the boats via VHF and the officer spread the word. Throngs of injured, civilians, and firefighters flocked to the waterfront.

With the help of Amico and Johansen on land, deckhands on each ferry facilitated boarding by hooking the boats' steel manoverboard ladders over the seawall railings so they sloped down the six feet or so to the deck at an easy 45-degree angle. Most passengers were able to scale the railings and descend the ladders to board the boats, but those who weren't received further assistance.

"If we had injured, we actually slid them down the ladder," Amico recalled. Even people in wheelchairs were able to board thanks to the cooperation of deckhands and other riders. "We had enough civilians to help. It was one of those things, no one stood back, whether civilian or employee," said Amico. Once a boat had injured aboard, the captain would turn and go. Another boat would glide in behind it. "At one point we were loading three or four boats at a time." And the people kept coming.

CHAPTER 6

"We're in the water!"

THAT MANHATTAN IS an island—a fact easily forgotten by modern day New Yorkers and visitors alike—was unmissable in the nineteenth century when, before bridges and tunnels, boats were the only means of travel on or off. The earliest crossings had been made by rowboats and periaugers (sailing craft with the option of oar-power supplementation). That held true until July 2, 1812, when Robert Fulton inaugurated steam-powered ferry service across the Hudson River between Manhattan and New Jersey aboard the double-ended steamboat *Jersey*. The new, oddly configured steamboat's regular 20-minute crossings launched the first-ever mechanically powered ferryboat service.

After that July day, steam-powered ferry services multiplied. By 1860, 11 different companies offered no fewer than 20 ferry routes to and from Manhattan. The early 1900s were perhaps the busiest years for New York harbor ferries, as steel-hull propeller boats became more common and city-run service was established. But as options for reaching the island expanded, ferry use began to dwindle.

In 1936, 117 ferryboats plied the waters of New York harbor—64 percent of them railroad-related or privately operated. But by 1975, just nine boats remained in operation, all of them run by the City of New York. Ridership decreased from 112.6 million in 1936 to about 20 million in 1975. As had happened with the Hudson and Manhattan (H&M) Railroad, the construction of bridges and vehicle tunnels drew commuters away from waterborne transit. In response to sharply reducing need, the last of the cross-Hudson ferries ended service in 1967, leaving the publicly owned Staten Island Ferry as the city's oldest, largest, and last standing ferry service.

For two decades, no cross-Hudson ferry existed. Then, in the mid-80s, the ridership battles between ferries and bridges and tunnels came full circle. With riverside developments cropping up along the Jersey waterfront, trucking company owner Arthur Imperatore recognized a new business opportunity. Bridges and tunnels were operating at peak capacity during commuter hours, which meant a waterborne transportation alternative—especially one offering a compelling four-minute crossing—could gain wide appeal. And so, in 1986, the *Port Imperial* ferry service was born, operating between Weehawken, New Jersey, and West Thirty-eighth Street, Manhattan.

Soon after, Imperatore's company submitted the winning bid to the Port Authority in response to its request for an operator to restore ferry service to the World Trade Center, via the World Financial Center terminal. Thus, more than 75 years after the idea of a public ferry service was inaugurated in New York City, a renewed trend toward private ferries began.

Later, Imperatore's expanding company was renamed New York Waterway, and by 1991, seven ferry routes carried more than 16,000 passengers daily. In 2000, ridership reached 32,000. The growing market soon attracted newcomers, including SeaStreak, sister company of Hoverspeed, an operator of English Channel ferries. On the morning of the eleventh, SeaStreak also lent its assets to the evacuation effort.

Tens of thousands of people were already streaming away from Manhattan on foot over bridges when, at 11:02 A.M., then-Mayor Rudolph Giuliani called upon everyone south of Canal Street to "get out." "Walk slowly and carefully," he said. "If you can't figure what else to do, just walk north." If this sounds vague, that's because it was. There was no plan. Even the planners had no plan for what was unfolding in Lower Manhattan.

"I'm a planner," explained U.S. Coast Guard Lieutenant Commander Kevin Gately in an internal oral history interview conducted eight months after the attacks. On September 11 he'd been a reservist for nearly 22 years, and his job with Activities New York's Waterways Management Division was planning periodic search and rescue and port readiness assessment exercises. The closest exercise approximation to the September 11 attacks, he conceded, was a series sponsored by the New Jersey Office of Emergency Management that had begun in the spring of 2000. The exercise had played out responses to a supposed terrorist attack on a Port Authority facility using a weapon of mass destruction.

When the events of that morning actually unfolded, he explained, "Our port security response was basically exactly what had been postulated in the exercise." But there was one significant difference: "The evacuation problem. That was completely unanticipated. That entire thing was improvised on the spot." The total evacuation of Lower Manhattan, he reiterated, had never been envisioned, was never dreamt of in our philosophy."

When the South Tower came down, Lieutenant Michael Day was on Staten Island standing in the Coast Guard Command Center, "riveted" to a big-screen television broadcasting CNN. "We got reports there were people congregating on the lower tip of Manhattan," he recalled. "That's when it really kicked in—when the first tower collapsed." VTS cameras showed people stacking up at the shoreline.

Boats of all kinds amassed along the water's edge, cramming their decks and interior spaces with evacuees, trying to deliver as many people off the island as possible. This unregulated effort raised Coast Guard concerns that overcrowding would cause problems on the water.

The acting captain of the port, Deputy Commander Patrick Harris, had very strategically perched himself on a high stool to coordinate the Activities New York response surrounded by representatives from all relevant departments. Haunted by visions of a Coast Guard boat that he'd seen "almost turning turtle" after being overloaded with refugees during the Cuban boat lift of 1980, Harris dispatched to "highly visible rallying points" a cadre of marine inspectors and investigators "with good strong command voices" who were knowledgeable about vessel capacities to ensure order and safety aboard ferries and to prevent other boat operators from loading unsafe numbers of passengers.

"We weren't as concerned about the fast ferries and those guys because they knew what they could do safely for passengers," he explained. "What we were really concerned with were the tugboats and the little private vessels—the guys that don't normally carry passengers."

Reports from on-scene mariners—the operators of tugs, small boats, ferries, and other vessels who'd made their way to Manhattan's shores almost immediately after the planes hit—continued to pour in, helping to augment what Coast Guard personnel at the VTS could see on-screen. But as the scale of the disaster compounded, the need for on-scene leadership became clear. Day would head out with a small team. But instead of using a Coast Guard vessel, Day decided to take Sandy Hook Pilot Andrew McGovern up on his offer for a boat.

The everyday job of the (currently 75 active) highly trained men and women of the Sandy Hook Pilots Association is to board all designated vessels as they enter or leave New York harbor, guiding them safely through the port. For more than 300 years, these local navigation experts have met schooners, steamships, and oceangoing vessels of all sorts at the entrance to the port to guide incoming ships across a series of shoals, called the Bar of Sandy Hook, that separate New York harbor from the Atlantic Ocean.

In 1694, even before New York became a state, the then-colony appointed the first local mariners as Sandy Hook Pilots, employing a term derived from the Dutch words *pijl* (pole) and *lood* (lead), which describe an early tool used to sound depths and chart waters. Initially these pilots operated independently, racing to be the first to reach a vessel and thereby secure the job of applying their local knowledge of tides, currents, shoals, and navigational hazards to guide a ship safely into port. In 1895, however, pilots from New York and New Jersey joined forces and established a regular working rotation. In 2015, the Sandy Hook Pilots made more than 10,000 trips aboard tankers, yachts, cargo, and cruise ships. They facilitated delivery of roughly 95 percent of all cargo entering the port.

All vessels longer than 100 feet that are flying a foreign flag or carrying foreign cargo are now legally required to have a licensed pilot aboard while traversing the harbor. This means that 24 hours a day, 365 days a year, in all weather conditions and port circumstances, Sandy Hook Pilots stand ready to board passenger liners, freighters, tankers, and other large ships on the open sea at the mouth of the harbor by stepping off the deck of a 53-foot aluminum launch onto the rungs of a ladder hung at midship. The work can be extremely dangerous. As the Sandy Hook Pilots Association president, Captain John Oldmixon, put it, "The chances of getting hurt are great, and the chances of you dying are significant if you mess up, or something goes wrong, or the ladder's not rigged right."

Securing the honor of serving in this precarious position is no easy feat. Until recently one had to know someone to land an apprenticeship. Times have changed somewhat, but still very few applicants are accepted into the association's rigorous five-year training program, which concludes with a four-day exam during which apprentices must reproduce from memory sections of nautical charts including every depth and buoy as well as each rock, reef, shoal, pipeline, and cable. Passing this test opens the door to an additional seven years of training as a deputy pilot before earning the designation (and salary) of full branch pilot. On September 11, the pilots not only knew where boats could safely tie up or load passengers, based on depths, currents, and hidden hazards, they also were familiar—from their work in the harbor every day—with the boats, the companies, and the mariners.

Surely their encyclopedic knowledge of the port and its people could be useful at a time like this. The question was, just how exactly? At this moment, no one was exactly sure. "There wasn't a preplanned response: This is what we do for two planes crashing into the towers," explained Day. Instead, "people were scurrying around" trying to figure out next steps.

Pilot McGovern had been driving to Manhattan for a Harbor Ops Committee meeting when the sight of the World Trade Center in flames made him reroute, beelining for the Fort Wadsworth Coast Guard Station where he knew he could get more information and offer up assistance. All our resources are at your disposal, McGovern had told Commander Daniel Ronan when he arrived.

One invaluable resource the pilots could provide was a mobile operating platform for coordinating the Coast Guard's on-scene response. Day agreed with McGovern that the 185-foot pilot boat *New York* would offer an ideal vessel for facilitating the maritime evacuation already under way, readily allowing for pilot and Coast Guard collaboration. Normally functioning as a combination command post and floating hotel, the highly maneuverable *New York* was well suited to staying on station for extended periods and offered a large wheelhouse with 360-degree views that was fully equipped with radar, radios, and other communications equipment.

In addition to the vessel itself, Day recognized the important contribution that the pilots' rich knowledge base, depth of experience, and strength of relationships with other mariners would bring to this emergency effort. He explained his highly unorthodox choice to join forces and use a non-Coast Guard asset as the base for Coast Guard activity in simple terms: the Sandy Hook Pilots "know that port like it's nobody's business."

Before Day set out to drive the short distance to the pilot station a few piers north, he pulled together paperwork, including nautical charts of the harbor and a copy of the plans that the Coast Guard had spent two years developing for the International Naval Review and Operation Sail (OpSail) event from the previous year.

On July 4, 2000, New York City had played host to a parade of tall sailing ships, naval vessels, yachts, and other ships from all around the world in what was believed to be the largest-ever port gathering, which included hundreds of security vessels as well as tens of thousands of pleasure craft. Managing the harbor traffic that day had presented VTS operators with the most demanding test in the center's history. Day thought the OpSail plans, which included medical and logistic staging procedures, might prove valuable as he set out across the harbor toward the unknown.

As it turned out, the OpSail event itself was already helping some mariners meet the challenges of the evacuation through the practical experience it had granted. During OpSail, instead of their usual 68 daily dockings at the company's own slips, New York Waterway captains moved 70,000 soldiers a day, often in unfamiliar territory. "We had to really pull into some bizarre places and offload the soldiers, [like] the sides of ships in a six-knot current," explained Port Captain Michael McPhillips. "I really think the captains gained a lot more experience doing that." Although the particulars of mariners' work on September 11 were unprecedented, OpSail preparations at least offered some guidance.

As he stood in the wheelhouse of the pilot boat *New York*, bound for Manhattan Island, Day had no concept that he and his team would soon be facilitating the largest waterborne evacuation in history. Or that he'd lose contact with his command.

At least I'll have a fighting chance in the river, Karen Lacey thought as she hovered at the edge of the seawall, preparing to jump. *I can swim. I can tread water. I'm an athlete. I'm not going to be stuck on the top of a building, hanging out of a window. I'm not going to be underground with a building collapsing on me. I'll have a fighting chance.*

From the time that she and Tammy Wiggs had left their other colleagues on the street corner a few blocks from the New York Stock Exchange, Lacey had made steering clear of buildings a priority. She hugged the shoreline as best she could on her way toward the World Financial Center ferry terminal, determined to get home to Hoboken. Now, as the pulverized tower plowed toward her, rolling like magma through the gaps between buildings and onto the waterfront plaza, the river promised salvation.

The sky went black as Lacey stood outside the rail. Chalky particles burned her throat. *I'm going in*, she decided, and plunged.

When she kicked herself back to the surface, she drew a choking breath and then buried her face back in the river, bobbing up and down half a dozen times before the sky changed from black to gray.

The drop to the river had been farther than she expected, and the current much stronger. She quickly kicked off her pumps, which were now making their way downstream, but she refused to drop the bag slung over her shoulder. Not only did it contain her wallet, the keys to her apartment, and tickets to that night's Broadway performance of *The Producers*, but the Coach tote was her second—a $400 investment she'd made to replace her first: a graduation present that had been ripped off at a bar. She clung to the seawall and to the bag, kicking to stay afloat.

Lacey heard Tammy Wiggs screaming, but Lacey said nothing.

"Now, I'm out to lunch. Now I'm spent. Now, I can't believe that we're here," she recalled. "I was trying to regroup. Maybe when the fog lifted, when the dust went away, I would get my bearings and start all over again but at that point . . ."

"We're in the water! We're in the water!" Wiggs yelled.

A gruff voice called back through the cloud, "Who's in the water?"

Lacey was farther north along the seawall than Wiggs and a good distance from the fireboat, so it took a while before she could see the boat, the ladder, or the bear of a man whose voice, low and booming, with a thick New York accent, served as a beacon in the murk.

The man calling down from the bow of fireboat *John D. McKean* might well have been firefighter Billy Gillman. It was Gillman's voice, at least, that rang out across the deck announcing to the rest of the *McKean* crew that two women were in the water. Engineer Gulmar Parga heard his call, as did wiper Greg Woods. Immediately Woods, an experienced lifeguard, grasped the peril of the situation. If either woman lost her grip on the concrete, the current would pull her downstream toward where the boat's hull banged against the seawall. She would be crushed.

Woods set off to collect the boat's Jacob's ladder, the best tool to bridge the gap between the river's surface and the deck. Meanwhile, someone on deck threw a rope over the side. Wiggs heard talk of a ladder but thought that the people on deck were saying they couldn't find one. With her fingertips still digging into a joint in the concrete seawall, Wiggs considered the line, thick as a Coke can with a loop at the end, as her only chance at rescue. But getting it into her hands wasn't going to be easy. The rope bobbed about three feet away. To reach it, she'd have to let go of the wall.

"All I could picture was getting sucked in between the boat and the seawall," she recalled. "It was all about timing . . . One, two, three, let go, push off, and you have one chance of grabbing this line because the current was ripping."

Wiggs let go. She pushed off. She caught the rope. Worried she didn't have much strength left, she decided to slip her legs through the loop to sit in it like a seat, and catch her breath. But as she kicked at it with her bare foot, the loop pulled through. Wiggs lost her grip on the line for a second but managed to snatch it back. She hollered up to the boat. "Give me more slack!"

But the people on the bow couldn't hear her clearly. Instead of feeding her more loose line they called out encouragements: "No ma'am, don't let go! Don't give up! Hold on."

Wiggs yelled up again, louder this time. "No. Give me some SLACK and I'll tie a knot that will hold!" Years later she'd laugh at herself recalling the moment. ("Here I am talking smack to the guy that's trying to save me.") The rope was the only thing keeping her from getting sucked into the gap between the wall and the boat's steel hull.

Finally, someone slacked out the line and Wiggs applied her sailor's knowledge to make a loop. Even as she treaded water, working with rope that was five times the size of any she'd ever before handled, Wiggs was able to tie a bowline knot that held. She put both legs through, sat in the loop, and the people on deck began yanking her out of the water.

Wiggs was most of the way up the side of the boat when, at last, the Jacob's ladder appeared over the side. She used it to step up the last few rungs to the cap rail. Now it was Karen Lacey's turn.

After she saw Wiggs swing her legs over the cap rail of the fireboat, Lacey inched her fingers down the concrete slab to get closer to the ladder, then pushed off to grab it. But even with her hands wrapped around the ladder's rope rails, she was still far from safe.

"The current was super strong and the ladder was super wiggly," Lacey recalled. "I can't get up the damned thing."

Although low water had hit the Battery at 8:50 A.M., the currents in this portion of the Hudson were determined by more than just the tide. Sometimes the current continued to pull downstream even as the flood tide began. Such were the conditions on the morning of September 11; many mariners reported a "ripping" ebb well into the ten o'clock hour.

As it dangled from the cap rail, the rope ladder with orange plastic rungs draped against the curve of the hull where it narrowed near the waterline. Simultaneously, the ebb pulled the lowest rungs downstream. The curled and bobbing ladder offered Lacey no leverage to gain her footing. Relying on just upper-body strength to haul herself out of the water proved impossible, and clinging to the ladder brought her even closer to the point where the hull slammed against the seawall. Lacey swayed a few feet from the point of impact.

"Come on. You can do it," the men called down to her.

"I can't," she yelled back. "So they're screaming at me: 'Drop the bag. Drop the bag.' And I was like, 'I'm not gonna.' 'Lady,' they said, 'drop the bag.'" *I can't believe I'm losing another one of these things*, she thought as she pulled the tote off her shoulder and let it go.

But bag or no bag, athlete or not, Lacey was tired. She couldn't lift herself up. Woods, the part-time lifeguard, jumped in after her.

"I saw his head bobbing in the water, and I was like, *Good idea!*" recalled marine engineer Gulmar Parga. He followed close behind, climbing down the pad eyes supporting the bow fender until he reached the water. He grabbed the ladder's rope side, pulling it taut.

Telecommunications specialist Rich Varela, the civilian who had boarded the boat only a few minutes earlier and was now bare-chested with a piece of shirt tied around his face, had been standing on the bow beside the lifeguard when he dove in. Now he hung over the side to help to stabilize the ladder from above.

Woods dove underwater. In a single swooping movement that left those watching from the deck somewhat awestruck, Woods hoisted Lacey on his shoulders and placed her feet on a ladder rung that Parga was bracing from the waterline and Varela was steadying from above. Once Lacey's feet were planted, she was able to climb enough rungs that Varela and others on deck could reach her arms and help her the rest of the way.

Finally out of the water, Lacey yanked her skirt back down over her shredded stockings, less embarrassed by the exposure than her inability to climb the ladder. Before her stood the man—wearing bunker pants, his glasses covered with a gray film—whose voice had called out through the cloud. Lacey gave him the first of the umpteen thank-yous she would deliver before deboarding in Jersey City. "I probably thanked him—conservatively—two thousand times. That's all I said the whole way up the ladder, down across the river, when I got off the boat was thank you."

CHAPTER 7

"Gray ghosts"

AT 10:28 A.M., the churning black smoke cloaking the top of the North Tower swelled, suddenly thicker and blacker. The upper floors of the superskyscraper cocked back, cracking open the west face of the building. For an instant the fires feasted, gorging themselves on the fresh supply of oxygen—orange tongues of flame licking out at the sky. And then, seemingly in slow motion, everything above crushed everything below.

Sean Kennedy watched it happen from the open-air helm of his thrill-ride speedboat, the *Chelsea Screamer*. "It started slow, breaking itself apart from the top," he explained in his languid Mississippi lilt. At one point, when the radio antenna that had stolen his focus seemed to pause in midair, he thought: *Don't fall! Don't fall! . . . Stop! Stop!* But the building didn't stop. It continued to cave. "It's coming down and there's a point where you can't say anything more. It's happened." Kennedy watched the spire plunge until it was engulfed in smoke and dust. He watched sections of the tower's iconic pinstripe columns slice through the southeast corner of 3 World Financial Center as metal fragments shot through the glass dome of the Winter Garden.

Kennedy, a longtime captain and charter-yacht owner who'd spent a lifetime on the water, was holding station just offshore of the Winter Garden, about 1,500 feet away from 1 World Trade Center. That particular distance was no accident. Kennedy had still been on land at 9:58 A.M., running down the West Side Highway on his way to the *Screamer*'s berth at Chelsea Piers, when 2 World Trade Center collapsed out of view. He didn't know how the tower had fallen, whether it had toppled "like a pine tree" or folded in on itself. Once he was out on the water he kept a careful distance.

On most days the *Chelsea Screamer* offered "splash and dash scenic adventures" around New York harbor. But today the 56-passenger tour boat was being used as a floating camera platform for a news agency crew looking to capture footage of the disaster. Three seconds before the building came down, Kennedy had been gaping in horror at the sight of bodies falling from the blazing North Tower. "A person!" he cried, his usual drawl overtaken by a tight, pinched screech. "Another person!"

"Holy sh—God! Mother of God!" shouted Kennedy's friend, crewmate, and fellow charter captain Greg Freitas, as the debris cloud mushroomed up, coursing riverine through every available opening, choking nearby buildings, branching out into tributaries, clogging every crevice, blanketing whole neighborhoods with the same stinging powder that the South Tower had unleashed a half hour earlier.

The building had twisted, unleashing a gale of wind—a burst of 55 million cubic feet of air—sending steel trusses screeching, glass erupting, columns crashing. Somehow, the noise made by all this destruction transported Kennedy back to his hometown in Biloxi, to the tracks where passing trains often stopped him on his way to school. While waiting to cross he'd listen to the clack of wheels. The sound that this mammoth building made as it crumbled— "kind of a continuous eruption of breakage"—reminded him of the thumps, clangs, and rattles of a string of rickety boxcars careening by at close range.

Minutes after the pulverized remnants of 1 World Trade Center settled out of the air along Manhattan's western shore, Kennedy caught sight of a Port Authority police officer waving him over to the shoreline. "Start evacuating people!" he ordered. "Anybody that can get over here we're gonna evacuate them." Kennedy complied. Though the bow of his speedboat sat well below the level of the loading platform, he nosed the vessel into a slip on the north side of New York Waterway's World Financial Center terminal to pick up anyone needing transport.

Spurred on by *Screamer* crew member Greg Freitas's indelicate urging, a handful of businessmen in white shirts and neckties bolted across the barge toward the speedboat. "Come on guys. Anybody coming? Get your ass over here, now. Now!" barked Freitas. "Come on. Come on! I want you to hold my hand and come on board. Get inside. Anybody else? Come on. Let's GO! Hold my hand. Get in. One at a time . . ."

Next Kennedy steered the boat south. Like the people trapped in the upper floors of the twin towers who had clustered at the windows, desperate for air, those caught in the avalanche of debris in Lower Manhattan fled to the water's edge, frantically trying to escape the choking cloud.

As he pulled into South Cove, a small rectangular notch cut out of Manhattan's western shore about 1,500 feet from the World Trade Center complex, Kennedy laid eyes on the people who'd been caught in that avalanche. "It looked like you had taken the ash from a fire and put it in a bucket and dumped it over them," Kennedy recalled. The dust clung to their clothes, to their bags and briefcases. People who had dressed in white that morning were now cloaked in a dark gray. When the *Chelsea Screamer* arrived, several police boats were already boarding passengers.

Kennedy pulled the *Screamer* into the cove, tying up alongside an NYPD launch, and passengers immediately began streaming aboard. Some evacuees stepping onto the bow of the police boat kept going to the vessel's recessed stern deck: a disoriented firefighter needed help donning a life preserver. A tall man in cargo shorts with dust-coated glasses and a towel around his neck pulled a dog on a leash. A shorter man wore black-and-white striped chef's pants.

Others, meanwhile, crossed from the bow of the launch onto the stern of the *Screamer*. Kennedy grabbed people's hands as they boarded, offering stability as they stepped across the ever-changing gap created by the two boats bobbing against each other. A young woman with long wavy blond hair toting a cat in a black purse cupped a cloth over her mouth. A round, middle-aged man in shorts with white hair and dark eyebrows carried a backpack. A preschooler in a yellow T-shirt clung tightly to a big woman in a baseball cap, his arms never loosening their grip around her neck.

At one point a man called down to the *Screamer* from Launch 3. "Does anybody have the baby?"

"Nobody has it," another man called back from the bow. A mother and her baby had been separated during boarding, Kennedy explained. "The baby went into the policeman's hand onto the police boat and she went into our boat," he said, adding that the two were quickly reunited.

More women continued to board with small children. A slender woman in a blue T-shirt passed a toddler to Kennedy before straddling the gap. A large woman in a pale yellow shirt handed over a chubby baby wearing short pants and squinting in the sun. Neither seemed to have been caught in the dust. Once the woman got her footing, she collected the infant and pulled him in tight.

Many people caught up in the unfolding catastrophe also had the duty of protecting children. Parents, teachers, daycare workers, and sitters were all forced to navigate their way through the danger while simultaneously soothing their frightened charges. Florence Fox, a 32-year-old nanny from Zambia, was one such person.

Nearly two hours earlier, at 8:45 A.M., in a Battery Park City townhouse on Albany Street a few blocks southwest of the South Tower, four-year-old Kitten was pestering her. "Let's go, Florence. Let's *go-o*," the girl mewled. Every Tuesday morning the two attended a story hour at the Borders bookstore in 5 World Trade Center, on the northeastern corner of the complex at the corner of Church and Vesey Streets. "Jeez, Kitten. Let me finish my bagel," Fox responded. "We're gonna go. We're gonna go."

Nannying came naturally to Fox, who'd helped raise her eight younger siblings. She had been caring for the girl she called "Kitten" since Kate Silverton was six weeks old. "I didn't look at Kitten like it was my job," she explained. "I treated her like my own child. I think she felt the same. She was very attached to me. She would tell me that she loved me."

Fox had just popped the last bite of bagel in her mouth when she heard an unfathomable sound. "I heard this noise. A big boom. Everything shook." She unlatched the front door of the house that opened directly onto the street. "I saw smoke. People were just screaming." Fox went inside and clicked on the television to find out what was going on, but quickly switched it back off. "I had to think of Kate. I didn't want her to see. I didn't want her to be freaked out." The phone rang. The girl's mother, Susan Silverton, had learned about the plane just moments after it hit and called to instruct Fox to stay inside until she returned home.

To explain to the child why they couldn't go to the bookstore, Fox made up a story. She distracted her with crayons and coloring books, arranging everything on a table before heading upstairs to watch the bedroom television out of view. Outside, the sounds of sirens and chaos grew louder, drawing Fox back downstairs to peek out the front door.

"All these people were dripping with blood, wailing and screaming. They were talking about people jumping." Someone yelled, *Another one! Another one!* and Fox heard a high-pitched peal streak through the sky. She slammed the door and ran back upstairs to the television. On-screen, a jet pierced through the south face of 2 World Trade Center just as the townhouse began to tremble in the wake of a percussive boom. Fox fell to her knees, shaking. *What do I do? I have this child.*

Back downstairs she tried to dial the girl's mother, but couldn't get through. Feeling the weight of a mother's trust, Fox was torn about whether to stay or go. Outside, the mayhem roared louder. "Florence, what's that noise?" There was no hiding the look on Fox's face. "She could see I was scared." For nearly an hour Fox struggled with her decision.

Then suddenly another massive concussion shook the house. *Oh my God*, thought Fox. *Now they've started bombing.* She opened the front door to see a wall of white debris moving at her "like a tsunami."

"Kitten, come over here now!" Fox screamed and the girl came running. She scooped her up in her arms and ran down the front steps. Neither had on shoes. "I don't even remember closing the door to the house," said Fox. Instantly they were plastered with debris. The child started to shiver and weep. "Don't cry," Fox told her. "It will be fine. Put your head in my breast and don't look."

Kitten buried her face against her nanny's chest, the woman's skin bare above her red scoop-neck T-shirt. Her small arms gripped Fox's shoulders, her little legs clenching at her waist as the woman fled barefoot through the street, trying to outrun the cloud. Like most people, Fox had no concept that the towers could fall. She assumed the city was being bombed. She didn't want to stop running but before long the powder burned her face and clouded up her eyes until she couldn't see. Time and again she paused for a second to wipe her eyes before continuing south on South End Avenue.

When at last she'd run far enough that the dust began to settle she could see that the neighborhood in which she'd worked for four years had been blanketed in a gray-white snow. She saw a man lying on the ground, wailing. She kicked him. "Get up!" she demanded. "Look at me," she said. "I have a child and I'm running. You have to get up. Let's go." She meant it as encouragement. And the prodding worked. The man got up and also began to run.

Up ahead Fox spotted people taking cover in a restaurant and she followed them inside. As she stood, still holding onto Kitten, people around her chattered, grasping at threads of information they hoped might explain what was unfolding outside the door. All conversation stopped short when a second rumble shook the earth. Fox closed her eyes. Screams erupted all around. "We thought they were bombs," Fox explained. "At this point people just lost it, just like wailing and crying." Kate bawled too. "Florence! Florence!" she called.

"I said, 'Kitten, it's okay. We'll be fine. It's okay.'" In that moment, staying strong for the little girl in her care was what propped Fox up to keep going. "She trusted me," the nanny explained. "Even with how she held onto me, she knew that I was going to take care of her."

With all the hysteria, Fox couldn't stay in the restaurant. "Up to this point, Kitten was scared but she wasn't subjected to other people screaming, and seeing fear." Fox knew she had to protect her from the crowd. "I cannot let her see this." Fox did not yet know how traumatizing this experience would wind up being for the little girl in her charge, but she was determined to do whatever she could to protect her.

Fox waited for the second wave of soot to settle before hitting the street once again. "I wiped my eyes. I went outside. And when I went outside I saw this little police boat."

"Gray ghosts." That was what the hoards of fraught, powder-plastered people looked like to Tony Sirvent as he nosed the police launch into South Cove for the first time. Known as a "cool customer," Sirvent was the NYPD Harbor Unit's senior pilot. For 30 years he had been bringing to his work both a commitment to serve and a no-nonsense, get-it-done approach. Today would be no different. That morning Sirvent's duty assignment was as pilot of the NYPD's 52-foot aluminum launch—known as Launch 9 or *Harbor Charlie*—stationed at the Harbor Unit's headquarters at the Brooklyn Army Terminal in Sunset Park, Brooklyn. That was where he had been when the planes hit.

Once known as the U.S. Army Military Ocean Terminal, the massive, 4-million-square-foot complex was built in only 17 months. From its completion in September of 1919 through World War II, the facility served as the nation's largest military supply base. Employing more than 20,000 military and civilian personnel, the site was the headquarters for the New York Port of Embarkation, a regional operation that moved 3.2 million troops and 37 million tons of military supplies to fronts across the globe.

Now, 56 years after World War II had ended, the facility where so many troops had shipped off would begin receiving evacuees fleeing a war zone right here at home, just five miles north in New York harbor. But Sirvent didn't yet know that when he and his three-person crew left the Harbor Unit base and went gunning toward Lower Manhattan. The launch had almost reached the southern tip of Governor's Island when the South Tower caved. Wind pulling the debris cloud to the southeast gave the officers a clear view of the devastation as they approached South Cove, where clusters of people had begun amassing along the water's edge.

As Sirvent steered into the notch, dust-covered people swarmed toward the boat, surging forward, pressing themselves against the railings, some scaling the wooden fences that separated the land from the water. "All right. Billy, get a line on those fences," Sirvent called to his newest crew member, Officer William Chartier, who'd joined the Harbor Unit only three months earlier.

At first Chartier didn't understand what the pilot had in mind. He tied a rope about the diameter of a Red Bull can onto one of the wooden fence posts, but it came undone right away. Sirvent fumed. "Here I am the new guy and Tony's on the flybridge and he's yelling at me: 'You're embarrassing us.'" Chartier felt pressure, not just from the flybridge, but from all the people "waiting to jump on this boat and get the hell out of Manhattan." This time he wrapped the line around a post and doubled it back to the boat, securing the loop end to the stem bitt—the post on the front of the boat that's designed to receive docking lines. When Sirvent backed down, hard, the force yanked the railing right off—which had been his intention all along. He was clearing the way for people to board more safely.

Sirvent's boss, the Harbor Unit's operational supervisor, was already in South Cove when *Launch 9* arrived. Now he protested that pulling down the fences was destroying property. "Hey Sarge, did you look around?" replied Sirvent. "Everything's destroyed already." And the crew continued to pull out the railings so that people boarding could step easily from the wooden ramps onto the bows of the boats—ferries, water taxis, and other vessels—converging on the scene.

"Once we did that we had people just streaming onto the boat," said NYPD Officer Tyrone Powell, on duty that day as the boat's navigator. "These dust-caked people, they were ready to go," he recalled. "Now we had like Noah's Ark. . . . We had everybody on that boat. We had animals. We had babies without parents. Everybody was covered in soot."

"They were handing us little children," confirmed Chartier. At first he was confused: *Where's the mother and father?* Later he learned that a nearby daycare center had been evacuated. "We took everybody. As many people as we possibly could fit onto the boat.... Ty and I actually tied down somebody in a wheelchair on the front." They set the wheelchair carrying a man, in his sixties, "more disabled than old," on the bow near the anchor and secured it to a handrail with ropes.

"Dogs, cats, everything came on this boat," said Powell. "That's why I call it Noah's Ark."

Sirvent worried about the nannies in particular. It bothered him that dozens of empty strollers had been abandoned along the water's edge after crew from other boats wouldn't permit them aboard. "I was thinking: *Wait a minute. They've got a two-year-old or one-year-old. They're only the nanny. They've got to tell people that they're somewhere in another state with their child and there's no phone service.*" So he instructed his crew to stow the strollers in the vessel's skiffs and then offload them with the passengers on the Jersey side.

Florence Fox didn't have a stroller. Just a four-year-old clinging to her, growing increasingly distressed. When Fox spotted the police boat, she decided it was her ticket to safety and rushed down to the water's edge. "You have to take me," she pleaded with the officers on board. "Look at her. I have a child." It's hard to say for sure, because no one was taking notes, but the launch she boarded may well have been Sirvent's *Harbor Charlie*.

After passing Kitten across, Fox boarded herself, stepping her bare feet down a winding stairwell to a small bunkroom. There sat another woman, tall and blond, holding a six- or eight-month-old baby. The baby was quiet, and the mother was too. Fox asked the woman if she had a cell phone. She wanted so much to call Kitten's mother to let her know they were safe. "I think she's the one that told me there's no service." Beyond that the two didn't speak.

It didn't take long for the boat to cross the river. Fox, who had never before been in New Jersey, found herself at a big office building among a group of Manhattan refugees and people handing out bottled water. She asked someone for a cell phone but then discovered that she couldn't recall Silverton's phone number. She had known the number by heart. But now, as would happen to so many people in shock that morning, she couldn't retrieve it.

She recognized several nannies from Battery Park. One held a baby about six months old. "She was just crying and the baby was crying and she was crying. I told her, 'You know, you have to take care of the baby. Look at the baby.'" Fox decided she couldn't stay there. "I didn't want Kitten to be subjected to that," she explained. Change in the little girl was already apparent in her expression and behavior. "I knew she was going to come out traumatized, and if I could save a little bit of that . . . Everything that I did I was thinking about Kitten." So she rinsed their hands and faces and started walking. *I'm going to find a hotel.* Carrying no wallet, and therefore no money or identification, Fox intended to call upon the kindness of the desk clerk. "I just wanted a room where I can wash Kitten, give her a bath, give her something to eat." As she walked, she asked residents for directions.

Time and time again, Sirvent and his crew maneuvered the police launch into South Cove to pick up passengers. "You know, God was with us, the maritime people," he said. What convinced him was how perfectly the tide lined up his bow with the wooden pier (now liberated of its railing) so that people could step easily onto the boat. The pilot and his crew proceeded to board around 100 people onto a boat rated to carry about 20. "We broke all kinds of Coast Guard regulations," Sirvent explained. "We had the cockpit full. We had people all on the outside deck. We had people actually sitting in the rowboat, the little skiff that was on top. Then we had people standing up in the flybridge with me."

Once every inch of usable space was occupied, Sirvent pulled away and headed due west, bound for an old ferry terminal in New Jersey's Liberty State Park. There, once again, a railing separated the water from the land. Passengers made their way over the metal gates and soon Sirvent backed the emptied launch out into the river. The Harbor Unit crew made 15 to 20 trips just like this. When clouds of dust left zero visibility, Sirvent switched on the radar and kept going, eyes red from the smoke, paper dust mask dangling around his neck.

As he pulled away from the ferry terminal after one trip, the pilot noticed that one of the passengers, a man in his forties carrying a tool bag that suggested he worked in some kind of construction, seemed to be examining the railing. Five minutes across, a few more for loading, and five minutes back. That was all the time the man had between Sirvent's trips across the river. But by the time the launch returned to the terminal, the railing had been disassembled and set off to the side so that the next load of people could just step off the boat. "This guy had the foresight to say, 'Hey, I have the ability. Let me take this time to help my fellow citizens by doing the right thing here.' And he did," That effort, that initiative, was "one of the little things" that Sirvent said has stuck with him.

During one of the police boat's many Hudson River crossings, Sirvent heard fighter jets roaring overhead. "When those [F-15s] were blowing down the Hudson River at about 400 miles an hour—I'm not saying I was crying because I don't cry, but—I had like tears in my eyes thinking about the might of this nation, and [that] there wasn't a damn thing that me or those jets could do to avoid this situation."

Although they couldn't stop the planes from crashing or buildings from falling, people from all quarters rose up and stepped forward to provide whatever assistance they could. All along New Jersey's North River waterfront, emergency medical technicians (EMTs) and paramedics, administrators and doctors, firefighters and police were working to establish and supply triage centers. Instrumental to their efforts were ordinary citizens helping every way they could. Personnel from nearby hospitals, medical centers, and emergency management offices worked with fire department and hazmat crews to establish makeshift facilities to decontaminate, assess, treat, and direct evacuees to different transit options. Their efforts were supported by the contributions made by employees of local businesses, among others.

One triage center was established directly across the river from the twin towers, in Jersey City's busy financial district just north of New York Waterway's Colgate Dock, off Paulus Hook. By about ten o'clock, the ad hoc treatment of injured people coming off the boats had become more organized. Soon, the waterfront plaza lay quilted with red, white, and yellow tarps. EMTs hovered, orbiting the injured who suffered from burns, breathing issues, broken bones, pelvic fractures, and lacerations, as well as emotional trauma.

Following mass casualty incident (MCI) protocols, the teams tagged patients based on the severity of their injuries. On victims with the most dire, yet survivable, injuries in need of the most urgent care, they used red tags. Yellow tags indicated patients in stable condition requiring hospital care but not in immediate grave danger. For the "walking wounded," and those with injuries so minor no doctor's care was necessary, they issued green or white tags, respectively. Black tags were reserved for the deceased and those unlikely to survive given available medical treatment. Later, a New Jersey EMS official would report that his group had triaged more than 1,000 patients in two hours based on how long it took for the team to exhaust the supply of 1,000 triage tags they had stored in their specialized MCI response truck.

Medical teams caring for Manhattan refugees flooding off the boats marshaled all available resources. Adjacent to the waterfront plaza, now laid out with first aid supplies and equipment, was a fenced-off construction site with a trailer. Emergency personnel asked construction workers to help them gain access to the air-conditioned trailer for asthmatics and others with breathing troubles. Within minutes the fence was plucked up and removed. Meanwhile, businesspeople from the surrounding office complexes rolled out office chairs that could be used to wheel the nonambulatory. They ripped down Venetian blinds and piled up the slats for use as splints. They yanked all the commercial first aid kits off the walls and brought out stacks of them.

Beyond medical needs, evacuees were focused on calling and getting home. With most cell service down, communications presented a huge challenge until employees of a nearby computer company set up two folding tables and ran out phone lines so that people could contact loved ones. Other people working in the area brought out janitorial uniforms for people whose clothes had been burned or destroyed. When EMTs asked for water, a cola company sent two trucks filled with 64-ounce bottles. A pharmacy brought a pallet of saline solution for people to use to rinse their contact lenses. In this manner, individuals' efforts to help—even in the smallest of ways—made important contributions to the evacuation as a whole.

Among the patients treated at the triage center was Kenneth Summers, the Blue Cross employee who'd been hit by a fireball bursting through the elevators into the lobby of 1 World Trade Center. "He was all frontal burns," recalled Mickie Slattery, a paramedic from the critical care team that transported him from the waterfront to the Jersey City Medical Center. Later, he was moved to the burn center at Saint Barnabas Medical Center in Livingston, New Jersey, where he remained for three weeks, undergoing four skin graft operations to his arms and hands. Not until the end of September would Summers's wife and daughter inform him that his workplace, Tower One, had collapsed, along with its twin.

At around 10:15 A.M., Tammy Wiggs stood barefoot on the bow of fireboat *John D. McKean,* her clothes (including the mesh-backed Merrill Lynch jacket she'd never removed) dripping Hudson River water into the eight inches of ashy gray soot that blanketed the deck. She began coughing so violently that she vomited, and once the heaving started it wouldn't let up. Soon Karen Lacey, also barefoot and dripping, joined her on the bow. Quickly thereafter, the Marine Division crew dropped lines and headed out across the Hudson, anxious to deliver their passengers, several gravely injured, to one of the triage centers that had sprouted up along the Jersey shores.

Lacey was surprised to see people she knew on board, including several colleagues. She had no idea how they'd gotten here, but now, as the ragtag group headed across the river, everyone exchanged what information they had. Lacey learned that buildings in Washington, D.C., had been hit and that several other unidentified planes were still in the air.

In the crush of trying to cover events that were heretofore inconceivable, journalists had unwittingly amplified the panic by reporting multiple false rumors. These rumors spread over the airwaves to televisions and radios as well as via the Blackberry devices commonly carried in the days before smart phones. Throughout the morning, broadcasters warned that multiple airliners were unaccounted for. Fears of another hijacked jumbo jet prompted the military to scramble F-15s to the airspace over the capital with orders to shoot down potentially dangerous planes. One suspicious aircraft was subsequently revealed to be a medevac helicopter.

At 10:23 A.M., the Associated Press reported that a car bomb had exploded outside the State Department. A minute later, 1010 WINS radio station in New York City announced an explosion at the U.S. Supreme Court. These and other reports of terrorist attacks, later revealed to be false, heightened the already skyrocketing anxiety of people nationwide. All the way out in Bloomington, Minnesota, officials evacuated and closed the Mall of America. By 10:53, for the first time since the 1973 Yom Kippur War, Defense Secretary Donald Rumsfeld ordered the U.S. Armed Forces placed at defense readiness condition (DEFCON) 3.

Before that, at about 20 minutes past ten o'clock, *McKean* pilot Jim Campanelli weighed possible docking options along the unfamiliar Jersey City shore-scape. At first he headed slightly north and west across the river, hoping to tie up at Harborside terminal. But when the 9.5-foot-draft boat ran aground in the eight-foot-deep channel, he backed out and headed south a quarter mile to another ferry landing at Paulus Hook, just south of the triage center.

Careful not to block the New York Waterway vessels offloading passengers at their regular berth, Campanelli pulled the *McKean* up to an adjacent dock. The current sucked the boat right against the dilapidated pier, making for a smooth landing. Though there were holes in the wood planking, the pier at least had functional bollards and cleats. The crew quickly made fast lines.

Marine engineer Parga was wrapping the bowline around a cleat on shore when he heard a horrified "Nooooooo!" erupt from the passengers on deck. He turned, still gripping the rope, to see the North Tower crumble, its antenna plunging straight down into the rolling dust.

From where she stood on the main deck of the *McKean,* Lacey had a clear, sickening view of the North Tower as it disintegrated before her eyes. "The first one was terror and fleeing," she recalled, "and the second one was, *Oh my God.* So many dead people. Just so many. I just couldn't even comprehend it. Now it wasn't personal safety, it was just sheer horror."

Wiggs, meanwhile, had been retching the whole way across the river. She was hanging over the side when a man interrupted her. "Ma'am, I know you probably don't want to be bothered right now," he began, "but for posterity's sake, I think you need to turn around." She raised her head to see the debris cloud mushrooming up. Not until some hours later would the notion of posterity become personal. That was when Wiggs discovered that the exterior columns that *Chelsea Screamer* Captain Sean Kennedy had watched slice through the corner of 3 World Financial Center had cut right through the section of the building where her sister worked.

Wiggs had dialed her sister Katherine "no short of 30 or 40 times" throughout her walk from the stock exchange to the Hudson. "Just hang up, dial again. Hang up, dial again." Sometimes she'd heard a busy signal, sometimes a recording announced that all circuits were busy. Katherine, eight days shy of her twenty-sixth birthday and 18 days away from her wedding, had fled quickly after the planes hit, but later returned to retrieve the bag she'd left under her desk containing the bridal undergarments she needed for a gown fitting scheduled later that day. *My mom's gonna kill me if I don't have these,* Katherine thought. Bag in hand, she managed to escape the building, again, and walk north to her apartment on the Upper East Side.

But Wiggs's college roommate had not been so lucky. The young woman, who had only begun working on the North Tower's eighty-ninth floor a short time earlier, perished. "The man who had me turn around," said Wiggs, "had me watching the building that killed my roommate."

Fireboat *McKean*'s crew, meanwhile, had just witnessed the deaths of hundreds of brother firefighters—friends and colleagues whose lives had flickered out in an instant. They were desperate to get back to the site. But first the crew had to offload passengers and cut through the two fences that stood between the evacuees and their deliverance.

To firefighters on the job, barriers to entry are little more than an everyday complication. Spotting the fences, marine engineer Parga grabbed a Halligan bar (a multipurpose forcible entry tool with an adze and pick end) and a gas-powered Partner saw. He summoned firefighter Tom Sullivan to help him cut holes through which the evacuees could reach the plaza—and, beyond it, the relative Avalon of Jersey City. While the *McKean* crew gnawed through the fences from the water side, Jersey police officers and firefighters chipped away from the land side. Soon they created openings big enough for both the ambulatory and the disabled.

As he had been doing since he first leapt aboard the fireboat, telecommunications specialist Rich Varela, shirtless, tattoos visible on his chest and arms, was helping fellow passengers. He was boosting less mobile people up and over the *McKean*'s cap rail onto the dock when he spotted a familiar face. The man whose lower leg had shattered when he jumped onto the stern deck was being transported off the boat in a chair. Varela thought he recognized him. *Was that the security guard who told us to leave the area beneath the pedestrian bridge?* Lending a hand to help lift the chair over the side, Varela felt a wave of gratitude that the man had convinced them to leave when he did.

When Lacey finally had her feet planted on the pier, she gave the man in the bunker pants and thick glasses, whose name she never learned, one last thank you. He reached out his arms and pulled her in for a bear hug. All she could think about, as she braced herself for the long walk home to Hoboken with no wallet, no keys, and no shoes, was that these firefighters were going back there. "Their mission had just gotten so much worse."

Varela harbored the same concern. When he heard the collective gasp that accompanied the crumbling of the North Tower he began to doubt that he would make it through the day. *All right*, he said to himself. *This could be it, man. This could be the last day of New York as we know it.* And then he thought of those firefighters on the boat.

They were frantic to offload passengers so they could return to the smoking ruins. "Our guys are there," Varela recalled them saying. "That's when I turned to Tom. I remember saying to him, 'I'm coming with you. You guys need help.' In my mind, I thought they were gonna be like, *No fucking way. You're not coming.*"

Instead, Sullivan said, "Let's go."

So Varela jumped back aboard fireboat *McKean*, joined by two other civilians, a Wall Streeter who yanked off his suit jacket and button-down before climbing aboard and an older gentleman who explained: "My son's in that building."

During the short trip across the river toward the wreckage, Varela's thoughts wavered between contemplating the dead and reconciling himself to the concept that this day might be his last. "It really felt like, *I might die today* . . . And I was okay with it." *These guys need help*, he thought. And that was it.

Chapter 8

"A sea of boats"

"Helplessness." That was the feeling consuming Lieutenant Michael Day on his approach to Manhattan shortly after both towers fell. Drawing closer to the smoke-choked Battery, he peered through binoculars at a foreign landscape. Lower Manhattan had become an achromatic world churning with dust and paper. The snow-like, debris-clogged gray air contrasted with the blue sky beyond the smoke. "You'd look behind you and it was a beautiful day. The weather was incredible," Day recalled. "And then looking at Manhattan . . ."

Desperate, ashy people—stacked 10 deep, maybe more—pressed up against the railings along the water's edge. Though "a sea of boats" had already rallied—tugs, tenders, ferries, and more, pushing into slips and against the seawall to rescue as many as they could—Day could tell that more boats were needed. Now, just before 10:45 a.m., the Coast Guard formalized the rescue work already under way by officially calling for a full-scale evacuation of Lower Manhattan.

"All available boats," Day began, issuing his first of many VHF marine-radio broadcasts summoning backup, "this is the United States Coast Guard aboard the pilot boat *New York*. Anyone wanting to help with the evacuation of Lower Manhattan report to Governors Island."

Back at Activities New York on Staten Island, the VTS had been issuing its own calls for mariners to respond. But Day said he never heard them. Clogged radio channels combined with the loss of one of the Coast Guard's main antennas from the top of 2 World Trade Center had left Day unable to reach his command.

So now, standing in the wheelhouse of the *New York*, he was the senior Coast Guard official on the scene of what would become the largest waterborne evacuation in history. "I didn't know I was going there to do an evacuation," Day conceded. "I was sent there, initially, to observe."

At 11:02 A.M., the Coast Guard's evacuation calls were echoed by New York City's then-Mayor Rudolph Giuliani. At this point, the evacuation mission grew exponentially. Now it was not only those caught in the immediate aftermath that needed transportation, but "everyone south of Canal Street." In fact workers were streaming out of buildings much farther north than Canal, all looking for a way home. While these people might not have been in immediate danger—though even that was unclear, given that the extent of the attacks was still unknown—they were still stranded, disoriented, and reeling. Such was the fate of office workers Chris Reetz and Chris Ryan, both sales associates at L90, an online advertising agency.

Oh shit, this is real, thought Reetz. Then, in the space of a breath, his fear mushroomed under the exacerbating sway of isolation. *I don't know anyone here.* Sure, the 25-year-old had coworker pals—people he could grab drinks with after work. Some of them were sitting right there with him in the office at West Twenty-third Street and Sixth Avenue when he heard the radio announce that a second plane had hit. But Reetz's roots were back in Detroit, where he'd been just days earlier. After a month-long layoff from the sales team at L90—a layoff he'd assumed would be permanent—he'd gotten called back to New York. His surprising rehire had felt like a lucky second chance, and this time Reetz had been determined to make it in New York City. Monday had been Reetz's first day back in the office.

Now it was Tuesday and e-mails and instant messages were popping up on computer screens all around the open-floor-plan office. "Where are you? Are you okay? What's going on?" Reetz and his colleagues, most of them 20-something transplants like him, typed reassurances to friends and family. Then a friend called from Michigan, frantic. He'd recently left a position at Cantor Fitzgerald, whose offices were located on the upper floors of 1 World Trade Center, and Reetz was the only person in Manhattan he was able to reach.

"What's going on? Can you see what's happening?"

"I have no idea. I just got out of a meeting." Slowly it clicked how many people Reetz and his friend knew who worked in those buildings. Reetz had their faces in mind when he rode the elevator to the top floor of his own building, two and a half miles away from the trade center. He and more than a dozen others stared through a bank of south-facing windows, mesmerized by the unobstructed view of the smoking towers, "almost paralyzed by our inability to look away." Then the South Tower fell.

I've got to get the hell out of here. Out of this building, thought Reetz. *I need to get somewhere safe*. But where was *safe* exactly? His inclination was to go home. But home wasn't in Manhattan. It wasn't even on the East Coast. The closest thing he had, an apartment in Hoboken, was across the Hudson River in New Jersey—unreachable now that, according to radio reports, the bridges and tunnels were on lockdown. *We're stranded here,* Reetz thought. *I have no idea where I'm going. I have no idea what to do.*

He was still standing, stunned, on the building's top floor, when a coworker rushed back down to the office, bursting through the glass doors from the elevator with tears streaming down her face.

"The tower fell. The tower fell," she bawled, making no attempt to conceal her tears. Chris Ryan was there in the office at the time. He was taken aback by the garishness of her horror, the unabashed grief. Her raw emotion—incongruous in the workplace—was so jarring that it took him a moment to grasp the words exiting the woman's mouth. Then their meaning came into focus. He wracked his brain to figure out what he should do next.

If they couldn't go home, at least they could get out of the office. In the wake of the South Tower's collapse, Reetz and Ryan accepted a coworker's invitation to walk to his apartment a few blocks away to learn what they could from the news. More than just colleagues, Reetz and Ryan were friends. They shared smoke breaks, after-work drinks, and sometimes the PATH train commute to or from Hoboken. After a few rounds of obligatory train small talk, the two had instituted a "no talking on the PATH" rule so they could each read their books instead. They'd wink, hold up their respective books, and dig in for the duration. Now that the PATH train shutdowns had left them both marooned in Manhattan they joined forces, finding solace in solidarity.

At the apartment, glued to the television, Reetz and Ryan watched as, over and over again, a plane plowed into the South Tower. Not until he saw the footage did Reetz fully grasp "the destructiveness, the truly evil act of flying a plane, with individuals on it, into a building, with people in it, with the explicit intent of killing."

Being stuck on an island with no way out had gotten Reetz thinking about the worst that could happen. *We're under attack. We don't know where else is going to be attacked. I want to get off of Manhattan.* This was paramount for him to feel safer. So when a reporter mentioned that ferries were crossing the Hudson to the Jersey side, Reetz was more than ready to flee. He and Ryan planned to meet up with Ryan's then-girlfriend and then head west to the water's edge.

As they stepped into the elevator on their way to street level, a man started railing. "I'm gonna kill any cab driver I see. I'm gonna kill any Muslim guy I see. I'm gonna kick their asses." *Is this what's going to happen next,* Reetz wondered, *people beating each other up on the streets?* Fear of riots and violence only heightened his resolve to get away. Just as they did on the PATH, Reetz and Ryan skipped the small talk on their walk across town. There wasn't much to say.

Aboard the Coast Guard's search and rescue boat *41497*, helmsman Carlos Perez was struggling with fractured comms. Despite his efforts to monitor Channels 16, 13, 12, 21, 22, 81, and 83, on both high- and very-high-frequency radios, communications were sketchy at best. Perez and his crew had been suffering through an extended radio silence when the evacuation call finally broke through. Soon after, they received orders to halt any pleasure craft between the Brooklyn Bridge and Chelsea Piers and to transport people from Manhattan to established triage centers. Perez shot, full throttle, toward the seawall surrounding Battery Park where terrorized civilians had fled toward the water's edge.

The distraught, disoriented, dust-cloaked people that Perez encountered there begged to be taken home to their families. Some were bleeding from cuts and scrapes, presumably sustained from falls on the sprint toward safety. "They did not care to hear that they were being taken to triages to be treated," Perez recalled. "It was as if every moment away from their families signified years. Rightfully so."

Every time Perez approached land, whether on the New York or New Jersey side, he had to jockey for position with other vessels, many of which were operating outside of their usual domains. Maneuvering was further complicated by the sheer volume of VHF radio traffic, which hindered communications between boats on the scene. "We had to physically come alongside vessels to communicate and coordinate the evacuation efforts," Perez recalled. Overall, he was impressed by how well the operators worked together despite restricted communication, taking turns pulling into slips or up to the seawall to board passengers.

Turn taking notwithstanding, reducing the logjams at passenger terminals was critical for ensuring maximum efficiency. Rerouting vessels that were ill suited to using ferry slips quickly became a Coast Guard priority in order to guarantee their availability for ferries.

From the pilot boat *New York*, Lieutenant Michael Day followed up his request for "all available boats" by asking that vessels report to marshaling stations to keep them from crowding dock facilities. The overwhelming response to the requests for mariner assistance further jammed the already congested VHF frequencies. "It was chaos," recalled Day. "Every channel you click to, people were screaming, 'Help! I need people over here! I've got someone hurt here!' Everyone was talking over everyone else." Initially, the team aboard the pilot boat asked each vessel to check in on Channel 73 with the name, size, draft, and passenger capacities of their boats so they could try to assign vessels to appropriate locations. The airwaves teemed with calls. "One of those tugboats would come in, 'Hey, where do you want us to go?' We'd say,

'Go to this pier.' . . . When you multiply that by all those different tugs and people checking in and giving them directions it was unwieldy."

Day said it was Coast Guard Chief Petty Officer Jaime Wilson who chimed in with a solution to help streamline operations. "Hey, Lieutenant," said Wilson, "why don't you just put them on a route going back and forth, from point A to point B, so you don't have to talk to them any more?" Day gratefully accepted the suggestion. In receiving, and then implementing, a suggestion from a subordinate, Day was following the example set by his command of having faith in the abilities of one's team.

From the outset, Acting Commander Patrick Harris had been demonstrating that leadership style from his perch at the Coast Guard station on Staten Island. Made more visible by sitting on that high stool, Harris was communicating that he was available for questions, approvals, and to bounce off ideas for action. "If people had any concerns they could look at me and realize: the boss is right there; I can do this," he explained.

> "People were used to acting on their own and they did a lot of that. . . . I did not have the depth of knowledge to micromanage. And I didn't pretend that I did. What I had was a depth of knowledge of the people I had working for me."

Harris trusted his people. And that trust trickled down.

So, with grease pencil on a whiteboard, Day, Wilson, other Coast Guard petty officers, and the half-dozen Sandy Hook Pilots crew aboard the *New York* collaborated to assign and map out drop-off and pick-up spots for tugs and other workboats that didn't have regular berths or equipment designed for easily on- and offloading passengers. To locate those spots, they referred to the detailed OpSail 2000 plan that Day had thought to grab on his way out the door. Soon evacuees were being funneled to distinct destinations based on where each vessel delivering them could dock. Deck crews began hanging spray-painted signs made of bedsheets and cardboard announcing their destinations. The makeshift ferries were now in service.

"We kind of thought of ourselves as air traffic control in that we were directing specific boats to specific locations throughout the city," recalled Day. Like his command, Day hoped the Coast Guard's presence could help expand, organize, and lend a bit of "legitimacy" to rescue efforts already under way.

As the primary regulatory body of the U.S. maritime industry, the Coast Guard devises and administers rules and standards governing every aspect of shipboard life, from the licensure and working conditions of mariners to the guidelines dictating vessel construction and equipment that affect all day-to-day operations of boat crews. The Coast Guard enforces the thousands of pages of rules contained in the multivolume Code of Federal Regulations (CFRs) that stipulate in exacting detail what is and isn't allowed on the water. As such, relationships between the Coast Guard and the maritime industry can, at times, become strained. Coasties (as they're called both affectionately and not so affectionately) don't only rescue mariners in trouble; they can also *cause* big trouble for boat crews caught in violation of the CFRs. Like restaurant health inspectors showing up unannounced, the sight of a Coastie approaching one's boat is not always a welcome sight, despite the vital services that the Coast Guard provides.

The choice that commanders made, in the midst of an unprecedented assault on the country that it was charged to defend, to allow mariners to bend rules in order to more readily assist their fellow citizens demonstrated a flexibility to adapt to the situation at hand that was critical to the success of the evacuation. So too was their choice to collaborate with mariners rather than control them.

For many mariners wishing to help but not yet engaged in evacuation efforts, the Coast Guard's calls for vessel assistance clarified their mission. Retired 1931 FDNY fireboat *John J. Harvey* was already bound for the trade center when pilot Huntley Gill heard the evacuation broadcast. The boat, which had been decommissioned and sold by New York City at a scrap auction in 1999, had no current FDNY affiliation or official duties. Nonetheless, several of the boat's civilian crew members had decided to bring it to the trade center in hopes that they could help. They hadn't heard the Coast Guard's earlier announcements that the Port of New York and New Jersey was closed to all nonessential traffic, and they were already pulling away from the boat's berth at Pier 63 Maritime at the foot of West Twenty-third Street when the cascade of debris from the second tower let loose. While the decommissioned fireboat was, technically speaking, now a recreational vessel, most of the Coast Guardsmen on patrol who were halting other pleasure craft and thwarting them from entering the harbor wouldn't have known that. Instead, the fireboat was likely considered to be what it looked like: just another municipal asset being called to the scene.

Less than 15 minutes later, Gill nudged the boat against the seawall just south of South Cove where a cluster of people stood along the shoreline near the Holocaust memorial. It looked to Chief Engineer Tim Ivory like people assembling outside during a fire drill. He called out to the crowd, explaining that the boat was here to transport them off the island. But before anyone would climb the ladder down from the top of the seawall to the deck they insisted on knowing where they were going.

"New Jersey," said Ivory. Nobody would board. Mariners from other vessels later shared similar stories about some people's reluctance to wind up out of state. Plenty of people in Lower Manhattan that morning lived farther up the island, or in Brooklyn, Queens, or elsewhere and they were likely wary of being stranded in a state they did not know, with no certainty about how or when they would be able to return home.

But when Gill announced over the public address system that the boat would head uptown instead, people poured on. Within minutes, 150 to 200 people had boarded and the fireboat was bound for Pier 40, a little more than a mile to the north. Before the boat made it past North Cove, however, a small police launch raced over, lights flashing. Officers called out to Ivory on deck, telling him to turn the boat around—that it was needed at the trade center site.

"We gotta get rid of these people," Ivory hollered back. "We'll go back after we get rid of all these people."

In that same moment, Gill heard a similar summons over VHF radio. From the wheelhouse of active-duty fireboat *McKean*, pilot Campanelli was hailing the decommissioned vessel. Firefighters needed it to pump water. *Harvey*, less than a month away from its seventieth birthday, was being called back into service to do the job for which the fireboat had been built. The order had come down not from top brass—so many of whom had been killed in the towers' collapse—but from Tom Whyte, an off-duty FDNY lieutenant.

———◦∞◦———

Whyte had heard about the attacks while at home in Hastings-on-Hudson, about 20 miles north of the World Trade Center. Although he wasn't scheduled for duty aboard fireboat *McKean* until later that evening, he'd quickly suited up, gathered his gear, collected his 21-foot pleasure boat, then shot south to the site, arriving not long after the *McKean* had returned to the Manhattan side from dropping off passengers in New Jersey.

Ushered over by firefighters waving them toward the shoreline, pilot Campanelli had docked against the seawall one block south of North Cove at the foot of Albany Street, less than 50 yards west of the Battery Park City townhouse from which nanny Florence Fox had recently fled.

Whyte explained his thoughts that morning as he sped downriver: "You just want to get in there [to help] any brother firemen, any civilians, anyone who's hurt, trapped. I figured there's gonna be body parts all over the place. And people stuck and trapped in cars . . . we just didn't know." Instead, what Whyte found, after heading ashore, having left his boat with neighbors who'd promised to run it back home, was "this crazy lunar landscape" and an eerie silence. "It was so quiet because everything that was there just got shut down. All the rigs were silent. It was just the weirdest. Red lights going from the batteries but no engines running. It was weird, weird. Scary weird. And the only thing that was on the ground was metal, dust, and paper. Everybody's personal lives you could pick up and read. You could trade somebody's stock. It was everything. It was just floating everywhere."

People fled to the water's edge, running until they ran out of land. © U.S. Coast Guard, photo by Brandon Brewer

Evacuees cluster along the waterfront just outside the fence surrounding the Coast Guard office building at the Battery, where Port Authority Commerce Director Richard Larrabee sought refuge after being caught in the first collapse. © U.S. Coast Guard, photo by Brandon Brewer

Fireboat *John D. McKean* carries evacuees to Jersey City. Passengers and crew aboard include Karen Lacey, Tammy Wiggs, Rich Varela, Bob Nussberger, Tom Sullivan, Gulmar Parga, and Jim Campanelli. © Ron Jeffers

Disembarking evacuees pass through the Jersey City triage center. Throughout the morning, New York Waterway ferries continue to offload passengers at the dock in the background, which is adjacent to the wooden pier (out of view) where fireboat *John D. McKean* delivered its passengers as the second tower fell. © Reena Rose Sibayan, *The Jersey Journal*

Passengers disembark just north of Paulus Hook in Jersey City. A ferry terminal receives smaller ferryboats and NYPD vessels, while larger boats, including fireboat *John D. McKean*, drop their passengers at the abandoned wooden pier on the left. On the shore, EMTs have begun laying out color-coded tarps to tend to the injured in their makeshift triage center. © 2001 New York City Police Department. All rights reserved.

Fireboat *John D. McKean* pumps water at the foot of Albany Street after its return to Manhattan, having dropped passengers at the Jersey City triage. © 2001 New York City Police Department. All rights reserved.

This aerial view, facing north, shows North Cove on the top left. Visible are Gateway Plaza (the tall, rectangular building just south of North Cove) and the World Financial Center buildings (with green domes and peaked roofs). Fireboat *John D. McKean* is tied up at the sea wall at the foot of Albany Street. Farther south, NYPD launches and other vessels can be seen nosing in to rescue evacuees at South Cove. © 2001 New York City Police Department. All rights reserved.

A stroller sits abandoned at the foot of Albany Street. In the background stands the town-house from which nanny Florence Fox fled with the four-year-old in her care. In the foreground lie hose lines that firefighter Tom Sullivan, IT specialist Rich Varela, and others stretched from fireboat *John D. McKean*. © JD Lock, johndlock.com

FDNY's Lieutenant Tom Whyte reports for duty at fireboat *John D. McKean* at the foot of Albany Street, shortly after the second tower collapsed. © Robert Deutsch, *USA Today*

Liberty State Park ferries and NYPD police boats pick up evacuees in South Cove. © 2001 New York City Police Department. All rights reserved.

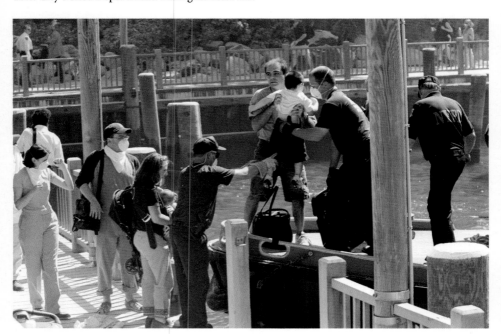

NYPD Harbor Unit vessels rescue adults and children from South Cove through gaps in the wooden railing created by Pilot Tony Sirvent and Officer William Chartier. © Matt Moyer

The view south from Manhattan shows tugs racing from Staten Island and New Jersey to the Battery. On the left is the southern tip of Governors Island. The Verrazano-Narrows Bridge is visible in the background. © 2001 New York City Police Department. All rights reserved.

Tugboats nose up to the sea wall along the southern tip of Manhattan to evacuate passengers. © Capt. Mike Littlefield

Hornbeck Offshore Transportation's ocean-going tug *Sea Service* becomes a makeshift ferry. Not since the earliest days of the tug industry were the same vessels used for both passenger ferrying and towing. © Capt. Mike Littlefield

A bed sheet announces Hoboken as the destination, and evacuees board by ladder across the bow of Moran Towing's tug *Turecamo Boys*. © U.S. Coast Guard, photo by Brandon Brewer

Fireboat *John J. Harvey* can be seen loading passengers just south of South Cove, adjacent to the Museum of Jewish Heritage. Farther south, New York Waterway ferries and tugboats load passengers from the sea wall in Robert F. Wagner Jr. Park as a U.S. Coast Guard vessel approaches. © 2001 New York City Police Department. All rights reserved.

The view north from above Governor's Island shows smoke-filled Lower Manhattan and the Battery at the southern tip of Manhattan. New Jersey is on the left. © 2001 New York City Police Department. All rights reserved.

Long lines for Circle Line, World Yacht, and New York Waterway boats at 38th St. on Manhattan's West Side extend for more than 30 blocks. Some wait for three hours to board a boat. © 2001 New York City Police Department. All rights reserved.

New York Waterway ferries and tour boat *Chelsea Screamer* (far right) evacuate passengers as the second tower collapses. The white tent covering the World Financial Center ferry terminal, where Pete Johansen has been helping to load passengers and dock builder Paul Amico is about to disembark, can be seen at the edge of the advancing dust cloud. Amico will soon establish a makeshift ferry terminal at the Downtown Boathouse pier (on the far left). © 2001 New York City Police Department. All rights reserved.

New York Waterway boats nose up to the sea wall at Robert F. Wagner Jr. Park, north of Pier A (with the green peaked roofed tower). Just out of view is the small dock where *Chelsea Screamer* Captain Sean Kennedy asked firefighters to break the lock so passengers can board. © 2001 New York City Police Department. All rights reserved.

Staten Island Ferry captains, including James Parese, make runs back and forth to Manhattan all day, evacuating more than 50,000 people. © 2001 New York City Police Department. All rights reserved.

The view southeast from above smoke-filled Lower Manhattan shows North Cove at the bottom center. Just south of North Cove, retired FDNY fireboat *John J. Harvey* has already begun pumping, as evidenced by the white spray pouring out from the boat's deck monitors that are not feeding hose lines. Farther south, active-duty fireboat *John D. McKean* is tied up at the seawall off Albany Street. Still farther south other workboats have nosed up to the seawall to receive passengers. © 2001 New York City Police Department. All rights reserved.

Still caked in dust, VIP Yacht Cruises' dinner boat *Excalibur*, with Captain Jerry Grandinetti at the helm, evacuates passengers from the Downtown Heliport. © Capt. Mike Littlefield

Moran Towing Corp. tug *Margaret Moran* delivers people from Manhattan across the East River. Red Hook Container Terminal's shipping cranes are visible in the background. © Capt. Mike Littlefield

A fleet of tugboats from different towing companies stand ready at the Battery. Several board passengers across ladders set up on their bows. © Capt. Mike Littlefield

1	Port Imperial	25	Albany Street townhouse
2	Pier 84 (Intrepid Sea, Air, and Space Museum)	26	Paulus Hook
		27	South Cove
3	Piers 81-83	28	Colgate Clock
4	Jacob K. Javits Convention Center	29	Robert F. Wagner Jr. Park
5	West 38th Street	30	Liberty Landing Marina
6	Lincoln Harbor	31	Morris Canal
7	Pier 63	32	Liberty State Park
8	Pier 62	33	Ellis Island
9	Pier 61	34	Liberty Island (Statue of Liberty)
10	Chelsea Piers	35	Pier A
11	Pier 60	36	Battery Park
12	Pier 59	37	The Battery
13	Pier 53	38	U.S. Coast Guard Offices
14	Hoboken	39	Whitehall Terminal (Staten Island Ferry)
15	Long Slip	40	Red Hook Container Terminal
16	Pier 40	41	New York Stock Exchange
17	Canal Street	42	Wall Street
18	Pier 26 (Downtown Boathouse)	43	Downtown Heliport
19	Chambers Street	44	Pier 11
20	Jersey City	45	South Street Seaport
21	World Financial Center Terminal	46	Brooklyn Bridge
22	PATH Tubes (formerly H&M Hudson Tubes)	47	Manhattan Bridge
		48	Brooklyn Navy Yard
23	North Cove	49	Empire State Building
24	Gateway Plaza		

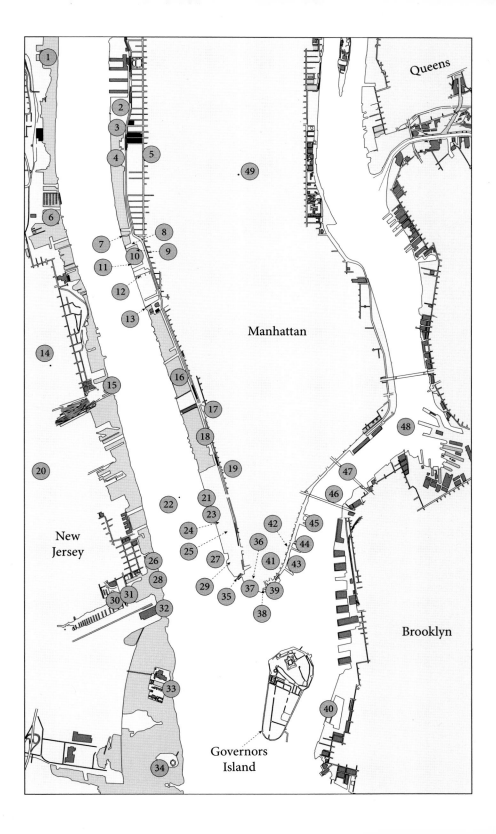

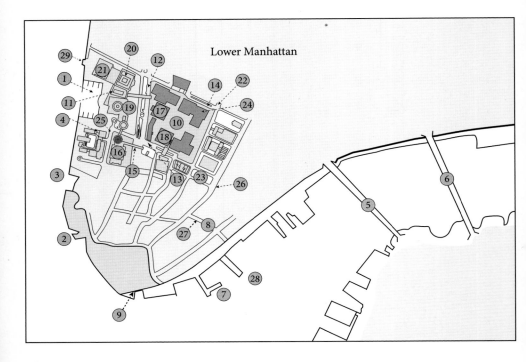

1	North Cove	17	1 World Trade Center (Tower One) (North Tower)
2	Pier A		
3	South Cove	18	2 World Trade Center (Tower Two) (South Tower)
4	Gateway Plaza (VIP Yacht Cruises offices and St. Joseph's Chapel on the first floor)		
		19	2 World Financial Center
5	Brooklyn Bridge	20	3 World Financial Center
6	Manhattan Bridge	21	4 World Financial Center
7	Downtown Heliport	22	Church Street
8	New York Stock Exchange	23	Zuccotti Park
9	U.S. Coast Guard Offices	24	5 World Trade Center (Borders Books and Music)
10	World Trade Center complex		
11	Winter Garden	25	South End Avenue
12	West Street	26	Broadway
13	Liberty Street	27	Wall Street
14	Vesey Street	28	Pier 11
15	South Bridge	29	World Financial Center Terminal
16	1 World Financial Center		

Whyte horseshoed around North Cove on his way to where fireboat *McKean* was tied up. When he arrived he was struck by "this crazed look" in the eyes of the crew. "They had a lot of hose lines out already.... The job was already under way."

Varela, the telecommunications specialist with no firefighting training, had been among those who'd stretched the hose. "I remember all hell breaking loose," Varela recalled of the boat's return to Manhattan after the second collapse. The deck was littered with debris. Strewn about in the six inches of gray-white powder lay clothing, briefcases, and shoes that the evacuated passengers had left behind in their haste. The shoreline swarmed with people—both firefighters and civilians. "People were rushing the boat," he said. The *McKean* crew waved them off. "Walk south! Walk south!" they commanded. Let the ferries and tugboats evacuate people. Fires raged all over the site and firefighters needed the fireboat's endless supply of river water.

While the crew handed off to land-based battalions every tool they could find—Halligans, crowbars, extra bottles for self-contained breathing apparatus—Sullivan tested a nearby fire hydrant and found it dry. The towers' collapse had burst water mains. River water would become essential to firefighting operations.

"What do you want me to do?" asked Varela.

"We're gonna start stretching hose," Sullivan responded. He and a *McKean* engineer had located an engine company on Albany Street a block or two away and the goal was to run enough hose to reach them. "Take the threaded end," continued Sullivan. "The threaded end goes toward the fire."

Varela followed every order: Take this line. Run it this way. Grab this line. Hook it up there. Run with it. Shirtless ever since he'd torn up his shirt for fellow passengers to use as makeshift dust masks, Varela now wore a life vest he'd gotten from the boat. "I looked like a bandito with this thing around my face," he later explained.

Before long the hose lines reached their destination, and fireboat *McKean* began pumping Hudson River water to land-based companies. Seeing that the job was already in progress, Whyte thought he'd head inland, closer to the site, to offer what help he could there. But, Sullivan addressed Whyte, now the highest-ranking officer, asking him to stay: "The officer for today, he's probably dead. He was there when the buildings came down. There's nobody here."

Whyte complied. "We all wanna be where the glory is, but everybody's gotta be where they're supposed to be." Not until another, regularly assigned officer arrived shortly thereafter did Whyte head inland. "I was making my way up Liberty Street and I was meeting these guys that were all banged up and bleeding. . . . Their eyes were big, wide open, you know? Just crazy, you know? They said, 'There's no water here anywhere.'" City buses, cabs, cars, and buildings blazed all around them.

While the *McKean*'s crew had successfully run hose lines on the southern edge of the site to feed a few tower ladders as well as the standpipes of nearby buildings, the whole north end still needed coverage. Out on the river, Whyte spotted the retired fireboat *Harvey*. Knowing the crew wouldn't have an FDNY-issued "handy talkie" radio but would carry a VHF marine radio, Whyte picked up his own handy talkie and called Campanelli, who could reach them by VHF. "Ask that old fireboat out there, the *Harvey*, ask them if they can come in and pump water. Because we got a spot here where there's nothing." Eyeing a nearby police boat, Whyte decided to reinforce the urgency with an in-person follow-up message. And so the police boat set out as well.

When Campanelli reported back that fireboat *Harvey* would return once they'd offloaded, Whyte followed up with another message: "Tell them to come in to Liberty Street [just south of North Cove] and I'll have some guys there." Next he rallied a crew of people—office workers, firefighters, and "guys in white shirts, carpenters, just a whole mishmash." He had them strip down smashed FDNY rigs searching for fittings, lengths of hose, anything usable. Because the 1931 fireboat was no longer in active service it did not carry aboard any of the accessories necessary to take advantage of its massive pumping capacity.

By the time the boat returned, Whyte was ready with supplies. "*Harvey* pulls up and I remember Timmy. He comes up out of the dungeon there," Whyte recalled, referring to then-Chief Engineer Ivory, who had just appeared from the below-decks engine room. "We just started heaving hose butts over the side."

Even after the towers had fallen, firefighters continued to battle multiple fires in surrounding buildings that on any normal day would have been considered big. "There was some good interior, structural firefighting going on," explained Whyte. But so much equipment had been lost in the avalanche of debris. Fireboats provided critical access to endless water when no other firefighting water could be found.

The minute he learned that the boat was being called to supply water, Ivory had begun preparing for pumping operations. From his four years as a volunteer with the Glen Rock Fire Department in New Jersey, Ivory knew that the ability to start and stop flow to supply lines was crucial for firefighting. So, with pipe wrench and sledge hammer in hand, he set about working to free up seized valves on the antique manifolds at the stern of the boat. Of 20 valves, only two functioned. But that meant he had at least *something* to work with.

Even before the boat docked against the seawall south of North Cove, Whyte hollered out the all-important question: "Do your pumps work?"

Ivory replied with a thumbs-up.

"Can you take lines?"

"Yeah."

"What do you need to make this work?"

"I've got about 300 gallons of fuel. I can pump for about 15 minutes." With that pronouncement, Ivory watched Whyte deflate. Days earlier the boat had returned from a run 120 miles up the Hudson River to Albany, New York, and back. "I was running on fumes," Ivory later explained.

Whyte radioed Campanelli once again. "*Harvey* needs fuel. Critical." And within 20 minutes, *Hayward*, an Army Corps of

Engineers drift collection vessel normally charged with clearing the waterways of flotsam and jetsam, arrived with diesel. Meanwhile Ivory, Whyte, and other firefighters struggled to overcome the next hurdle: the decommissioned fireboat's three-and-a-half-inch valves didn't match the FDNY's current standard three-inch hose. *Harvey* had none of the necessary adapters. Hearing this, Whyte directed *McKean*'s engineers to gather up as many fittings as they could find.

While Ivory waited for adapters, it occurred to him that the fire department's three-inch hoses *would* fit into some of the boat's deck monitors—the large brass guns that firefighters traditionally used to direct giant powerful streams of river water directly onto blazes by raising, lowering, and pivoting the gun tip. *They've gotta be New York thread*, Ivory reasoned. The boat was too far from the fires to spray water directly so attaching hose lines that could be fed by the vessel's powerful pumps was the only option. Ivory's creative problem solving led to an unorthodox use of the deck monitors: he removed the gun tips and screwed on the hose lines directly, creating a makeshift manifold.

As he prepared to fire up the pumps, Ivory worried about how much water pressure would be lost through the deck guns to which he hadn't attached hose lines. Because he couldn't gate them down, those deck guns would run freely, diminishing the much-needed pressure through unfettered streams. Then he hatched an idea. He screwed off the nozzles on those unneeded deck guns and stuffed a full, capped water bottle into each one. Those water bottles served to thwart the flow and thus diverted additional pressure to where it was needed: to the hose lines running long distances through the dust to firefighters on land.

When Whyte returned he was impressed by Ivory's innovative solution. "This is very important," he told Ivory. "Get a pen and paper. You have to keep tabs on where this stuff is going." Using sidewalk chalk that some kids had left behind after a past boat ride, Ivory numbered the deck guns and drew a schematic on the side of fireboat *Harvey*'s wheelhouse, creating a guide to keep track of where each hose ended up so that he could stop the flow to individual lines as needed. One hose, recalled Whyte, wound its way to Fifteen Truck's tower ladder, which then channeled river water into 10 or 15 hand lines that firefighters were using to confront blazes.

When the Coast Guard issued the call for "all available boats," a fleet of tugboats were among those who responded. Aerial photos taken by the NYPD Aviation Unit reveal ribbons of quickwater unfurling behind the flotilla of tugs shooting across the harbor toward the tip of Manhattan. Some were small harbor units. Others were huge oceangoing vessels. And on an ordinary day in 2001, these boats would have had no business carrying passengers.

For modern mariners, ferrying and towing don't mix. Tugs have a very specific job to do—they push and pull barges and other vessels. The notion of them being used to move civilians instead would have been anathema for most vessel operators in the harbor that day. But repurposing tugs as ferries was actually harking back to a much earlier time in the harbor when the demarcations between passenger and cargo transport were not yet so clear.

When Robert Fulton debuted his newfangled *North River*, the first commercially viable steamboat running on the Hudson in 1807, observers described the odd-looking contraption—150 feet long, 13 feet wide, and drawing just 2 feet of water—as a "devil going up the river in a sawmill" and "a monster moving on the waters defying the winds and tide, and breathing flames and smoke." At the dawn of the industrial revolution, only a handful of steam engines existed in the country and the devices remained foreign to most people. Alien as it was then, Fulton's successful experimental vessel established the dawn of the steam era, awing passengers with its ability to operate with a consistency that defied the vagaries of winds and currents.

Soon steam power increasingly replaced wind as the prime mover of both passengers and cargo. By 1850, more than 150 steamboats traveled up and down the Hudson, ferrying as many as 2 million passengers. Steam power also revolutionized the movement of goods over water, and the business of towing began, albeit informally.

Early on, steam ferryboats would sometimes divert from their scheduled routes (often with a full load of passengers aboard) to assist sailing vessels into or out of port. Such tows typically only occurred under extenuating circumstances, like when weather, damage, or crew troubles meant that another boat really needed assistance. In time, however, as shipping vessels doubled in size (making maneuvering more challenging), and the average cargo size multiplied (making offshore offloading more time-consuming and costly), towing became more commonplace.

After the City of Brooklyn passed an 1832 ordinance forbidding steam ferries from ditching their regular runs for more lucrative ship-assist work, the first New York vessel designed specifically and exclusively for towing and shipping-assistance arrived on the scene: the *Hercules*. At 116 feet long and 192 tons, this substantial towboat was made available 24 hours a day. Alas, before long, limited call for the vessel's services led its owners to convert operations to passenger ferrying. Over time, however, as more and more freight traveled by barge and all those barges needed towing, New York harbor developed a thriving industry of boats whose sole purpose was to move other boats.

By the late 1800s, the screw propeller-driven, "model bow" tug, built in the iconic tugboat shape still prevalent among children's toys, was the most common kind of tug in the harbor. Every aspect of the design of these ships was driven by their function. From hull shape to rudder position to propeller number and location, the main concerns were for maximum power, maneuverability, and efficiency.

Among other concerns was the prevalence of boiler explosions, ship fires, and other deadly catastrophes on the water—many due to negligence or greed. These eventually led to ever-stricter safety requirements. Meeting those standards, which would ultimately be enforced by the Coast Guard, called for specialized construction. By the end of the twentieth century, the diesel engine had come to replace steam, entirely dominating the world of commercial marine propulsion.

Tugs facilitated the development of critical industries, including the railroads, by enabling the transport of fuels (first coal and later oil) as well as industrial chemicals. But soon, the shift to containerization from break-bulk cargo, as well as reductions in the transport of Hudson River Valley commodities that had once been industry mainstays (including stone, sand, cement, brick, ice, and coal), led to a significant downturn in New York harbor's towing industry. By the early 2000s, the fleet of New York tugs had dropped from its 1929 peak of 800-plus to fewer than 300 boats.

Although overall tug traffic had been vastly reduced by 2001, a handful of tug companies still provided services essential to the everyday functioning of the port including ship docking and the transport of petroleum products, chemicals, and dry bulk materials like coal. On the morning of September 11, these tug companies and crews dropped their usual duties to provide another essential service—one for which they had long since lost authorization or design capability: carrying passengers off the island of Manhattan.

Virtually every towing company in and around New York, whether large or small, committed resources to assist with the evacuation of Lower Manhattan. Since tugs tend to work jobs in combinations of no more than a few boats at a time, when Staten Island Ferry Captain James Parese saw "a sea of tugboats" bound for Manhattan, it made an impression. "I couldn't believe the amount of tugs."

Among the tug companies that sprang into action was Reinauer Transportation. Marine engineer Glenn Dorhn had been supervising repair work aboard the tug *Stephen Scott Reinauer* when a welder ran into the bunkroom that was being replated, yelling. "I don't speak Spanish, but I can tell when somebody's excited," he explained. "The boat could have been sinking, far as I know." So he followed the welders racing onto the dock. The Reinauer

Transportation yard, located on the north shore of Staten Island, afforded clear views of Lower Manhattan. From across the harbor, Dorhn could see what was causing all the commotion.

In the Reinauer offices, meanwhile, the higher ups were trying to figure out what action they could take, discussing crew safety and liability issues along with questions about exactly what kind of help their boats, which were completely ill suited (not to mention unauthorized) for ferrying passengers, could provide. After the South Tower collapsed, they summoned the tug captains to the conference room to ask for volunteers. All four captains agreed to go to the trade center site and offer whatever assistance they could. Reinauer's port captain and safety director, Ken Peterson, who'd been a mariner in New York harbor for more than a quarter century, would head up the fleet.

Before setting out, the crews ransacked the company warehouses and boat lockers, gathering up water, blankets, towels, respirators, extra life jackets, oxygen, first aid kits—anything that might be useful—though they had no idea what to expect, and no clear mission. Long gone were the days of tow boats designed to readily accommodate passengers, but crews aboard the four tugs responding—the *Franklin Reinauer, John Reinauer, Morgan Reinauer*, and *Janice Anne Reinauer*—figured they might take civilians on board and cleared the decks of as much towing equipment as they could to reduce tripping hazards. With no assurances that attacks on New York City had ended, tug crews set out from the safety of Staten Island to offer what aid they could. Many did so out of a sense of decency, a sense of duty, and in an effort to quell the unbearable sense of helplessness haunting them.

When the request came down for Reinauer crews to head over, Dorhn grabbed a life jacket and stepped onto the dock. The *Franklin* and *Janice Anne* had already pulled out and the *John* was backing away so he jumped aboard the *Morgan* just before deckhands released the head line securing the boat to the dock. "I was just going to go to help in any way I could," said Dorhn. Anything would be better than standing on the dock "watching the smoke rise."

From the wheelhouse of the *Franklin Reinauer*, Peterson noticed tugs from other companies also charging toward the southern tip of Manhattan. When he saw throngs of people pressed up against seawall railings, soot-covered and bloodied, wailing, frantic, and desperate to flee, the mission became clear. He radioed the Coast Guard for permission to pull up to the Battery and pick up passengers. With permission granted, Peterson got on VHF Channel 13 and began issuing instructions to other tugs: "Everybody line up, bows-on, and get ready to take people."

Here, the tugboats had entered uncharted waters. Sure, they'd managed crew changes in dicey situations that demanded that crews clamber up the bow fendering or jump over open waters. But those leaps had been taken by seasoned mariners, not everyday civilians, most of whom had never even seen a tugboat this close-up. As soon as the *Franklin* reached the seawall, Peterson climbed ashore to help orchestrate the makeshift ferry service, instructing captains to load 100 or so passengers each time, knowing full well that most boats carried no more than about nine life jackets.

The tugs set about delivering people wherever they wanted to go, to any location with enough water for the deep-draft boats. Some headed for Jersey City, others left for the Brooklyn Navy Yard, some went to Port Newark, New Jersey, or Staten Island. "I don't remember all the destinations," Peterson explained, recalling several people who wanted to go to Long Island. "Well, take 'em out to where you know that they can get a cab or a bus or something," he instructed one Reinauer captain. "We did go to the airport," he said. "We did go to Rikers Island [the site of New York City's main jail complex, in the East River between Queens and the Bronx]." Once the crowds thinned out a bit, no minimum passenger quota was necessary to justify a trip. "It didn't matter. The day was just..." Peterson explained, trailing off. "If it was six people or four people that was enough."

Crews spray-painted bedsheets as destination signs. "Hoboken," read one, its pronouncement granting some semblance of surety to stranded people who had already that morning been through enough mayhem. Once aboard, deckhands greeted the powder-plastered people with towels, sheets, blankets, and water so they could wipe their faces. Their efforts to provide some measure of comfort made all the difference.

When the captain of the *Morgan Reinauer* first nosed the tug up to the seawall lining the Battery, the park at the southern tip of Manhattan Island was caked in white powder. The leaves on the trees were coated with a chalky dust. A smell hung in the thick, soot-filled air that reminded Dorhn of "burning furniture and hair and debris." One after another, the crew helped people climb the tug's bow fender and directed them to the stern, which offered the most deck space for people to congregate.

Deckhands managed to lift a man, and then his wheelchair, up and onto the deck. They assisted an older gentleman who explained that recent knee surgery had left him struggling to run from the cloud. When a tall, slender woman in her sixties with dyed black hair and all-black clothing boarded, Dorhn joked with her about his favorite singer, "Man In Black" Johnny Cash. Soon two mentally challenged men in their fifties boarded, their caretaker following close behind. Twins, dressed alike in dark sweat suits, the men were dust-covered head to toe, all except for their eyes. The places where they'd wiped away the grime left behind matching raccoon-mask streaks.

After escorting all these people to the back deck, Dorhn made his way to the bow just as deckhands hoisted up a little girl in a pink dress. "This was the first *child* child that I had seen," he explained. To Dorhn she looked to be about five years old. "She was fairly clean, so I don't know if she was inside or around a corner when the towers came down and the dust was flying," he said. The deckhands didn't set her on her feet; instead they handed her to Dorhn, who held the girl in his arms.

Figuring that she probably wasn't alone, Dorhn waited. Next over the side was a woman in her twenties, thin, with sandy blond hair, and very noticeably very pregnant in summer clothes. It struck Dorhn as odd, as she climbed aboard, that she didn't address the little girl or register any hesitation about a stranger holding her daughter. "She was just like a zombie" with a "totally blank stare. I said, 'Just follow me.' And she did."

"My name is Glenn," he said to the girl, wanting to reassure her. "What's your name?"

"Natalie," she said. "Where are we going?"

"The sun is shining and the weather is warm so we are going for a little boat ride," he replied, leading the two toward the stern bitts at the back of the boat.

"Set down here," he said to the mother, indicating a coil of eight-inch line that sat about a foot and a half off the deck, figuring it would be a bit softer than the hard steel. He placed the girl beside her.

"Is there anything I can do to help make you more comfortable?"

The woman sat mute. The little girl pointed at the smoke column rising from where the towers once stood. "My daddy works there."

With that the mother crumpled. "You could have turned on a spigot and it would have been less water than the tears coming down her face," Dorhn recalled. "I'll give you some tissues, some napkins," he said, rushing off to the galley, biting his lip to maintain his composure.

Upon his return, before being pulled away to help board other passengers, he knelt beside them, offering the only other thing he could: "I'll say a little prayer for your husband."

Soon after, its decks filled with at least 60 passengers, the tug backed away, bound for Hoboken. The passengers disembarked and the crew returned to the seawall to rescue still more. According to Peterson, 27 tugboats evacuated more than 4,000 people from the Battery over the course of four hours, even as an additional five tugs circled back and forth evacuating people from Pier 11 on the East River side of the island.

CHAPTER 9

"I need a boat."

A LITTLE FARTHER upriver on the west side of Manhattan, Paul Amico was finally right where he felt he should be—where the land meets the water. The iron worker and dock builder was straddling a section of two-foot railing atop the seawall north of North Cove when the second tower collapsed. He had been loading injured passengers with a group of firefighters. But as debris rained down, Amico leapt behind a short, three-foot-high knee wall along the esplanade and pressed his body flat against the pavement. Not knowing whether the building would fall straight down like the first or topple forward, landing right on top of him, Amico covered his head and waited.

When he looked up again, the first thing he saw was a row of empty firefighters' boots standing at the seawall with bunker pants folded down around them. "It was a very eerie sight," Amico recalled. "Eight or ten pair of these just standing there without people in them." He quickly processed that when the firefighters saw the building coming down they had flipped off their suspenders, jumped out of their boots, and run. Soon enough they returned, stepped back into their gear, and resumed working. But the image stuck with him.

Reports of broken gas lines in the area had led police and firefighters to shut down all loading operations at and around the ferry terminal. Now evacuees had to walk farther and farther north. The Waterway ferries continued pulling into the seawall to meet them, but docking conditions were hardly ideal. The captains struggled to hold the boats' rounded bows tight against the concrete wall despite the current pushing them broadside. Instead of lining up passengers for loading at distances of 40 or 50 feet apart, Amico had to grant each boat a wider berth, allowing a 150-foot spread to account for north-south movement of boats pushed by the current. He realized it would be much better if the boats could load taking the current at the stern. This thought led Amico to the Downtown Boathouse.

"That was my kayak club. I was a member there," Amico explained. "I knew we had about nine feet at low water. The ferries draw seven." This meant the water was deep enough for the ferries to dock safely. "So I just moved the whole operation down to the boathouse." Located in a cement building on the north side of Pier 26, about a half-mile north of North Cove, the Downtown Boathouse ran public kayaking programs. Amico held a key to the building, and he knew where the acetylene torch was stashed. Within 40 minutes after the second tower fell, Amico created a loading spot much safer than the seawall by cutting through a chain-link fence on the boathouse's cement pier. Here the ferries were better positioned to handle the current, nosing in to the northwest corner. The ferries' bows rested almost level with the pier, allowing passengers to board straight across, no climbing required.

To maintain contact with the police department, Amico asked the sergeant he'd been working with to follow him to the new berthing area. Waterway Operations Director Pete Johansen joined him as well, continuing his crowd-control efforts. Amico communicated this new loading arrangement to the ferry captains by marine radio, then flagged them over by waving his bright orange life vest. "Look for me," he'd say over the radio. "I'll be at the end of the pier."

"Paul, where are you?"

"I'm waving the life jacket."

At one point an FBI agent approached him. "I need a boat. I have 20-plus people I have to get over to Brooklyn now."

So Amico called in a boat. "You're taking these guys to Brooklyn," he instructed, no questions asked.

By late morning an armada of different vessels, from dinner yachts to tour boats to tugs, had responded to the disaster unfolding in Lower Manhattan. The captain of one tug contacted Amico by radio to ask how he could help. At the time a big timber was floating near Pier 26, posing a navigation hazard and occupying valuable frontage. Amico asked the captain to put a line on the wood and tie it off somewhere, to which he gladly complied. His assistance cleared enough space for an additional boat to load passengers from the pier.

On the morning of September 11, mariners brought specialized capacities to all manner of diverse tasks. As Amico explained, "To work on the water, you have to be a problem solver." That on-the-fly, solution-oriented approach—along with the steadfast willingness to help—proved invaluable on this grim and forbidding day.

Just before 10:30 A.M., VIP Yacht Cruises Captain Jerry Grandinetti was back in North Cove after dropping off the firefighters. He was tying his rigid inflatable to the dock when suddenly, once again, the sky went black. He sought shelter in the dinner boat *Lexington* until the worst of the debris fall subsided. Smoke and dust hung so thick in the air that he could not see up. He had no inkling that this fourth "explosion" was a second 110-story building pancaking to the ground. Nonetheless, he decided it was time to pull at least one of VIP's boats off the dock.

From working the previous night's dinner cruise, Grandinetti knew the *Excalibur* had fuel. But he'd have to clean about four or five inches of dust off the sloped windshield to see out. His attempts with a squeegee failed under the weight of all that powder, so he headed to the engine room to start the fire pump. With the fire hose he sprayed down the windshield of the *Excalibur*, cleaned the windows on the *Royal Princess*, and hosed off the *Lexington*'s canvas canopy.

Then, stepping through the muddy mess that the wet soot had made on the deck of the *Excalibur*, he headed back down below to fire up the main engines, three 12-71 Detroit diesels. Grandinetti had spent years engineering aboard fishing boats, so getting engines running posed no problem. What would be tricky, however, was backing the 132-foot yacht out of the slip with no help handling deck lines. He'd have to release the bow and stern lines and single up the spring line (the line that led diagonally from the boat to the pier to keep it from moving too far forward or back), then run up to the wheelhouse and put the boat in gear, then run back down to release the spring line, and sprint back up to the wheelhouse, leaving enough time to grab the wheel before the boat collided with other vessels or the surrounding docks. "I figured I could handle it," he said. But before he freed the last line an NYPD officer yelled down to him from the World Financial Center plaza.

"Hey, you gotta get out of here!" he hollered, alerting Grandinetti to suspected gas leaks.

"I'm trying to get out of here," Grandinetti yelled back. "I'm trying to get this boat out of here."

"You can get that boat outta here?"

"Yeah, I can get the boat out of here, but once I get out to the river I'm fucked because I've got no crew."

"You need a crew?"

"That'd help."

"Give me five minutes," said the cop, pulling out his radio.

Minutes later a small NYPD Harbor Unit boat pulled in and two officers boarded the *Excalibur*. "You can carry a lot of people," one officer said.

"One hundred forty-nine passengers," Grandinetti confirmed.

"We're gonna help with the evacuation."

"No problem," replied Grandinetti. "You just let me know what you wanna do." This was the first he had heard of any formal evacuation.

Grandinetti pulled the dinner boat out into the Hudson and headed south. After rounding the Battery to the East River, he saw that Pier 11 was congested with other boats. So the captain continued about 1,000 feet to the south to the Downtown Heliport. With tugboats nosing all along the outside edge of the L-shaped pier, Grandinetti steered for the inside, hoping he wouldn't hit bottom with the six-foot-draft vessel that did not, after all, belong to him. Despite having taken the unorthodox step of volunteering his company's vessel to the rescue effort, the seasoned captain didn't worry about the owner's disapproval. "He knew me well enough. He wouldn't mind," Grandinetti said. "Whatever I was doing was okay."

And what he was doing at this moment was trying to figure out how to secure a boat to a pier with no fendering, no tie-up, and just a ladder leading down into the water. He told the officers to pull a spring line through the rungs of the ladder—"the only thing bolted to concrete"—and tie it back to a cleat on the boat. Then he adjusted the throttle to put sufficient tension on the rope—enough to pull the boat in tight so that people could jump a narrow foot-and-a-half gap, but not so much that it would yank off the ladder.

"All right, start bringing them on," said Grandinetti, asking the officers to try to keep track of how many people climbed on. They lost count, but after a good number of people, many dust-covered, had boarded, and it was time to pull out, the officers asked Grandinetti to head to the old Brooklyn Army Terminal, the location of the Harbor Unit's headquarters.

Once there, along the stretch of waterfront where warships once lined long piers waiting for resupply, and soldiers (including Elvis Presley) bid farewell to their homeland before deployment overseas, Grandinetti maneuvered his way to an open spot amid the tugs and other dinner boats already dropping off Manhattan refugees. From the outside wing-station controls, he spotted the VIP owner's son, deckhand Bob Haywood Jr., running down the dock waving his arms. "I don't know why he was on the pier," Grandinetti said. "He lives in that area. He was probably trying to figure out a way to get up to the [North Cove] marina and then he saw me pull in."

"Let him on," Grandinetti called down to the people manning the ramp. Now at least he would have an actual VIP crew member aboard. No sooner had Grandinetti offloaded his passengers than police officers on the dock asked the captain if he'd be willing to shuttle some medical personnel to Manhattan. Their destination? The World Trade Center site. And so, as quickly as he'd made it off the island, Grandinetti was headed back. It would end up being just one of many, many trips he'd make aboard the dinner boat that would serve as a ferryboat, a taxi service, and then a supply boat over the course of the day and through the night.

As it happened, fellow VIP captain Dennis Miano was also in Brooklyn that morning with Grandinetti's bosses, Mark Phillips and Bob Haywood Sr. They had all been trying to figure out what they could do to help when an image of the *Excalibur*, Grandinetti at the helm, appeared on a nearby television screen. "We couldn't believe the cheek," recalled Miano. "How the hell did the boat get out of the damn dock?" But, just as Grandinetti had expected, his bosses supported him. Soon after, they too headed to the waterfront, boarded a small boat, and set out for Manhattan.

"I need a boat." Spirit Cruises Operations Director Greg Hanchrow had driven nearly 20 miles farther north, away from New York City, to reach the Petersen Boat Yard & Marina in Upper Nyack, New York. Now he was petitioning a long-time friend for a speedboat that he hoped would deliver him to the Spirit Cruises fleet berthed in Manhattan.

"I can't give you a boat," replied the friend.

"Give me a fucking break. Give me a fucking boat," said Hanchrow. "A fast boat."

Finally the friend complied. To this day Hanchrow doesn't know whose boat he purloined, but the big bow rider with two 125-horsepower outboards was, indeed, fast. "The thing did like 40 fucking knots, man. It was like a car." Speed was important now that he had wound up about 30 miles away from Chelsea Piers.

Next Hanchrow needed a crew. He knew most of the Spirit Cruises crews were stuck outside the city and he'd at least need deckhands to pull the boats off the dock. "Dude, guys, come with me," he said to two men working in the boatyard. When he explained his plan, the men agreed. Hanchrow hadn't yet heard anything about an evacuation. Instead what compelled him was "a captain's sense of duty." He was worried about the vessels in his charge.

He pulled the go-fast boat out of Petersen's and shot south. "I hooked it up," he said. "We were fucking booking." Less than an hour later they sped beneath the George Washington Bridge and a Coast Guard boat charged after them, catching up only when Hanchrow pulled into Chelsea Piers. After he showed them his driver's license and business card and explained his intention to move the Spirit boats, they left him alone. "I probably didn't fit the kind of profile that they were looking for," he explained. So he tied up the speedboat and got to work.

The three dinner boats, *Spirit of New York*, *Spirit of New Jersey*, and *Spirit of the Hudson*, were tied up in their usual berths on the north side of Pier 61. There, Hanchrow met up with two Spirit captains who'd made it to the pier earlier that morning, plus the general manager. The first priority, all agreed, was to get the boats away from Manhattan and head upriver. So they began firing up engines.

Before they could pull off the dock a Chelsea Piers security guard came running over, explaining that people were swarming north, trying to get off the island. Could the Spirit boats help? Hanchrow walked around the north side of the blue siding-clad Chelsea Piers building to see hoards of people trudging up all six lanes of the West Side Highway. "I'd never seen anything like it in my life. It was like if Yankee Stadium had evacuated. 50,000 . . . 100,000 people." Distraught residents, commuters, and visitors fleeing the toxic fog overhanging Lower Manhattan streamed up the main artery of Manhattan's west side, winding their way north along the shoreline in search of a way out.

Chelsea Piers has long been a place where people have found themselves at life-changing crossroads and pivotal moments in history. Stretching between Twelfth and Twenty-second Streets along the Hudson River, the nine large docks, once edged by a row of grand buildings adorned with pink granite facades, were designed to berth the largest ocean liners of the day. One of these, the famed *Titanic*, had been scheduled to dock at Pier 59 on April 16, 1912, at the conclusion of its maiden voyage. Instead, only its lifeboats arrived courtesy of the Cunard liner *Carpathia*, which had rescued 675 of the ship's 2,200 passengers from the icy ocean and delivered them back to terra firma via the Chelsea Piers.

Three years later, in May 1915, the Piers became the last place that passengers boarding the luxury liner *Lusitania* bound for England would ever set foot on land. Off the coast of Ireland, a German U-boat torpedoed the oceanliner killing 1,198 people, 124 of them American, in an attack that prompted the United States's entry into World War I. Soon, soldiers sent off to fight that war boarded troop carriers berthed at the same piers that served variously, over the decades, as the city's leading passenger ship terminal, boarding and debarking celebrities and steerage-class immigrants alike; a debarkation point for soldiers shipping overseas, this time to fight in World War II; and, in the late fifties and early sixties, a cargo terminal.

Following the decline and then abandonment of much of Manhattan's active waterfront in the 1970s and '80s, the Chelsea Piers were turned over in the early nineties to a private company for redevelopment. The new Chelsea Piers Sports and Entertainment Complex opened in stages. By 1996 the complex filled four piers, 59 through 62. Today, the buildings house film and television production facilities; events spaces; sports fields, courts, and training centers; ice rinks; restaurants; a mile-long waterfront esplanade; and more. Along with a marina for private vessels, the piers themselves provide docking facilities for charter and dinner-cruise boats, among them Spirit Cruises.

Not until he stood gaping at the crowds filing up the West Side Highway did Hanchrow notice that both World Trade Center towers had vanished from the skyline. His new mission became instantly clear. *If we're going to take a 600-passenger boat out*, he thought, *let's take these people with us.* Hanchrow used the marine radio to contact his good friend Michael McPhillips, still stationed in the wheelhouse of New York Waterway ferryboat *George Washington*, which was continuing to ferry passengers across the river to New Jersey. "Mike, what do you got up at Thirty-eighth Street?"

"All hell is breaking loose," came McPhillips's reply. "The Coast Guard has ordered all boats to evacuate the island."

Hearing the pronouncement that a full-scale evacuation was now under way eliminated any doubts Hanchrow might have had about filling the Spirit Cruises dinner boats with passengers. In some respects these vessels, designed to load and offload large numbers of people quickly and efficiently, couldn't have been more perfect for this mission. Trouble was, where could the captains disembark so many people on run after run to New Jersey? Long before Hanchrow had entered the dinner-boat world, he'd put in plenty of years working on vessels that moved things instead of people. This sensibility informed his thinking. "It's like a container ship," explained Hanchrow. "If you can't get containers off the dock what's the good of having a big container ship?"

And so, leveraging his decades in the harbor, Hanchrow started making phone calls. He called the general manager of the Lincoln Harbor Yacht Club in New Jersey, Gerard Rokosz, whom he'd known for years, and learned that the New York Circle Line Sightseeing Yachts had already begun ferrying passengers to that location.

Harbor veteran Circle Line had been running New York City sightseeing cruises for decades, offering its signature narrated tours with views of the Manhattan skyline, the Statue of Liberty, Ellis Island, all three bridges connecting Manhattan to Brooklyn, Yankee Stadium, and the Empire State Building, among other New York City landmarks. Founded in April 1945 (once the fuel shortages of World War II had eased) the company had gotten its start when a handful of Irish boaters pooled their resources—cash and vessels—to launch a business providing three-hour yacht cruises that circumnavigated Manhattan Island. Over the years, Circle Line managed to buy out or overtake most rivals, holding onto and expanding its corner of the water-based sightseeing market. In 2001, business was still going strong.

But as anxieties mounted on this Tuesday morning, Circle Line's operators stepped out of their regular orbits in favor of the straight lines that would provide the shortest distance between two points. By 10:15 A.M., people flooding north on foot from Lower Manhattan started piling up at Pier 84, at the foot of West Forty-second Street, prompting the company to begin shuttling passengers—600 at a time, aboard its three largest tour boats—on continuous trips across the Hudson to Weehawken. Word spread that the company was offering free passage to New Jersey and by midday, aerial photographs showed thousands of people standing in lines that continued for more than 30 blocks, from the West Fifties south to West Twenty-third Street. Some waited for three hours to board a boat. By nightfall, six Circle Line boats had transported about 30,000 people.

Once Hanchrow learned that Circle Line was already moving passengers, he dialed a friend at the company to talk through logistics. It seemed to Hanchrow that Lincoln Harbor, located about a mile across the river in Weehawken, New Jersey—a straight shot northwest from Chelsea Piers—was the best drop-off option. So, shortly after the second tower came down, the Spirit Cruises captains launched their own shuttle service from Pier 61 to Lincoln Harbor with Hanchrow at the helm of the *Spirit of New Jersey* while his colleagues captained the 145-foot *Spirit of the Hudson* and the 192-foot *Spirit of New York*. Deckhands aboard the three boats used clickers to make sure they stuck to their certified capacities: 600 on *New York*, 575 on *New Jersey*, and 425 on *Hudson*. "It didn't make sense to go over capacity," explained fellow Spirit Captain Daniel Scarnecchia. "What would we have done, got there 10 minutes quicker? The gain wasn't going to overpower the potential loss." As the Chelsea Piers guards and auxiliary police helped funnel foot traffic to the pier, the boats loaded up and dropped lines for their first seven-minute runs across the river.

Next the captains had to figure out how to tie up safely without use of their usual docking equipment or facilities. While the other boats made their way toward the harbor's breakwater, Hanchrow hoped to dock at "a tiny spot" along the south side of the seawall.

Upon his approach he discovered that it was occupied by a Reinauer tug. He watched as an evacuee crawled her way to land, rung by rung, across a horizontal ladder suspended over the water from the tug to the seawall. This was hardly a safe maneuver, nor an efficient way to offload passengers. Having put in plenty of years as a tug captain, working for Reinauer among other companies, Hanchrow didn't hesitate before picking up the radio mic to bark out orders at the tug's captain. "Dude, you've gotta get that fucking thing outta there."

"We're doing what we have to do to evacuate people," the captain responded.

"You've got 12 people on that boat and they can't get off because of your fendering," said Hanchrow. "Get that fucking thing out of there. I've got 600 people I'm going to get off in 10 minutes. It's taking you five minutes to get one person off. You've gotta move. Nose up to me and get your people on my boat." And so the captain did.

Then Hanchrow set about docking the 175-foot vessel at a spot that "looks like a postage stamp when you're trying to land the *Spirit of New Jersey* on it." The gangway they'd brought with them was too long for the narrow section of seawall, so crew members scouted out another shorter section of gangway from the assortment that different charter operators stored there. Hanchrow radioed his friend, VIP Yacht Cruises Captain Dennis Miano, to ask if he knew whose gangway it was. He said he didn't know. "Just take it." So Hanchrow and his crew rigged it to be able to disembark passengers from the main deck cargo door. Their plan worked well, at least until the tide went out.

Even if they weren't quite accustomed to ferry duty, the Spirit Cruises captains were certainly in the habit of carrying passengers. Unlike at the Battery where proximity to the tower collapses and the resultant smoke and debris conditions caused outright panic, the people who'd stepped into long lines winding through the parking lot between Pier 62 and Pier 63 remained orderly. Even at its peak, when the lines ran up the West Side, people stayed calm, just grateful for a ride home. Each time the three dinner boats loaded 1,600 passengers and set back out across the Hudson it was clear that they were making a dent. By day's end the Spirit Cruises vessels had delivered approximately 8,000 people off Manhattan Island.

Lincoln Harbor Yacht Club's general manager, Gerard Rokosz, couldn't stand there any longer watching the towers burn. Seconds after the first plane hit, he'd heard the call that sailboat *Ventura* Captain Pat Harris had made to the Coast Guard. He'd looked out the window of the club's corner office to see fire licking out from the north face of Tower One. After the second plane hit he'd stared, like everyone around him, at the rolling smoke until he "got tired of looking at it and hearing all the chatter on the radio." He contacted the Coast Guard to share the best resource he had to offer: "We have a pier if you need it."

Not long after, he received a call from someone at Circle Line asking if its boats could disembark passengers at his marina. Rokosz granted permission. With the arrival of the first Circle Line boat minutes later, and the first Spirit Cruises dinner boats not long thereafter, Lincoln Harbor's evacuation depot service had begun.

The Lincoln Harbor dockmaster, meanwhile, had been scrambling to get to the marina from Bayonne since he'd heard about the second plane. That Tuesday was Janer Vazquez's day off, but even the street closures that prevented him from getting to Lincoln Harbor by car didn't deter him from heading in. He'd driven as far as his mother's place, on the Bayonne–Jersey City border, where he grabbed his nephew's bike. Then the five-foot-11-inch-tall, 41-year-old man pedaled the child-sized bike two miles to Liberty Landing, where he knew a friend of a friend who had a boat. The boat owner agreed to run Vazquez the four miles upriver to his workplace in his 27-foot Sea Ray, but before they reached the mouth of the Morris Canal a Coast Guard patrol boat halted them.

Once Vazquez explained that his boss, Rokosz, had called him in to help with the evacuation efforts the Coast Guard personnel escorted the Sea Ray to the marina. Vazquez arrived just as the first boat pulled in. For hours thereafter he scurried back and forth along the 500-foot dock making fast and letting go lines and helping people disembark as boat after boat dropped its passengers.

Evacuees arrived by the thousands, many of them with no clear sense of where to go next. The managers of the nearby Academy Bus Company, however, recognized that they had access to an essential resource for facilitating a waterborne evacuation. Rokosz received a call: we've got buses trying to get over to you but the roads are closed. So Rokosz called the Weehawken Police Department and spoke with a sergeant he knew. Before long Academy buses were shuttling thousands of people from the marina to either the Hoboken train station or Giants Stadium, a centrally located landmark from which people could arrange a pickup or other transportation. The flood of evacuees reminded Rokosz of "the retreat from Stalingrad." They wanted to know where the buses were taking them, he recalled: "'New Jersey,' I told them. 'It's not that big a state . . . You're not far from anything here. We'll get you home. Someone will pick you up. We're taking you to the stadium. Call somebody and get picked up.'"

Soon after passengers began lining up to await Spirit boat transport, the hundreds of people in queue in the parking lot just north of Chelsea Piers became thousands. Seeing the crowds swell, the operator of a neighboring pier stepped up to offer his own docking facility for use in the evacuation.

Back in the 1980s, when Manhattan's western shore was deemed a wretched, fetid eyesore rather than valuable frontage, an electrical contractor from Seattle named John Krevey had gone looking for a cheap place to set up shop. He'd wound up renting space in a dirty, rat-infested paper warehouse adjacent to the all-but-abandoned Chelsea Piers. Back in Seattle, waterfront property had always been the most expensive. It struck Krevey as odd that here "in the great Port of New York" riverside real estate was the cheapest. Such was the harbor's state of decline. The no-man's-land along the water's edge allowed a bit of freewheeling on Krevey's part. Recognizing the potential of the small stretch of shoreline behind the warehouse, he decided to expand his empire.

So, maximizing the 40 feet of frontage included in his lease, Krevey cleverly docked the short end of a 360-foot-long rectangular barge at the water's edge, thereby extending his domain a full 14,400 square feet into the Hudson River. Having installed the barge—a former car float once used to transport cargo-laden freight cars across the Hudson from New Jersey for offloading—he dubbed it Pier 63 Maritime, a public-access pier that quickly became an unparalleled "old salt" destination. A New York City attraction with no hint of a "touristy" feel, the barge soon became a tiki bar/restaurant/arts space. The scrappy little Shangri-La, where visitors could watch a performance, take in a sunset, learn to paddle a kayak or outrigger canoe, explore historic boats, hop on a dinner cruise, or dance away a warm summer night, appealed to gritty boat people and young partiers alike.

On September 11, Pier 63 Maritime called out like a beacon, drawing countless mariners and community members to this outpost that had come to feel like a home. The magnetic pull to this spot helped galvanize a critical component of the maritime evacuation by offering another route off the island. After all, a successful evacuation depends on docks as well as boats.

From a motorboat on the Hudson, Krevey had watched as the South Tower spewed a torrent of soot that smothered his Battery Park City neighborhood. Then he headed north to the foot of West Twenty-third Street. As the boat traveled upriver, Krevey watched throngs of anguished, ashy people surge up the West Side Highway. He wondered how he could help.

Back at Pier 63 Maritime, he spoke by phone to the operators of the 600-passenger boat *Horizon*, which had previously scheduled a charter to depart from the pier at two o'clock that afternoon. "I talked to them about possibly being involved in a relief effort to get some of the people out," Krevey recalled. Soon enough, "they pulled in and we started forming lines." He quickly set his plan in motion, transforming the pier and the parking lot into a makeshift ferry terminal.

"Bruce, I need you." When retired administrative project manager for the Metropolitan Transportation Authority, and Pier 63 regular, Bruce Rosenkrantz arrived shortly before 11 A.M., Krevey immediately put him to work.

"You've got me," Rosenkrantz replied, relieved to be given a job—some concrete task to perform in the midst of all this havoc. He set out for the parking lot on his assigned mission, but when he registered the masses of people in the long lines, he faltered. *I don't want to cause a stampede*, thought Rosenkrantz, before returning to Krevey to voice his concerns.

This time Krevey escorted him back out with what was, Rosenkrantz explained, a "perfect Krevey" solution: a roll of red paper tickets. They spotted an auxiliary police officer, handed him the roll, and asked him to distribute the tickets to people needing a ride across the river to New Jersey so they could begin forming a new line. Bifurcating the river of evacuees eager to get off the island, Krevey had created a distributary channel—another way out.

Soon more Pier 63 regulars arrived—many of whom had headed there instinctively, drawn by the sense of the pier as a sanctuary. Among them was John Doswell, a waterfront consultant and maritime event producer, who quickly set to work helping to mold the lines into switchbacks so that more people could occupy the limited footprint of the parking lot in front of Basketball City. Accustomed to organizing people waiting to party, not evacuate, Doswell, Krevey, and other volunteers nevertheless applied their crowd-control skills to this new situation. They rigged up blue ropes tied to white plastic lawn chairs to organize the line. Krevey stepped among the people, holding his favorite megaphone up to his lips so he could answer individuals' questions for the benefit of all within earshot.

By happenstance, the subway train that former Fiduciary staffer Bonnie Aldinger had hurtled herself into in the wake of the second impact was an uptown train. That determined the course of the rest of her day. Instead of heading home to Brooklyn, she stepped off at Twenty-third Street, where she knew of a diner with a television. There, over a glass of orange juice that she hoped would cool her nerves, news reports filled in the blanks about the fate betiding her former colleagues on the North Tower's ninety-seventh floor.

She also learned that the bridges and tunnels connecting Manhattan Island to the rest of the world were being shut down. With home no longer an easy option, Aldinger set out for the next best place: the public-access pier that *felt* like home, Pier 63. In the wake of her layoff, she'd spent much of the summer paddling with the Manhattan Kayak Company, where she was both an instructor and a partner. The relief she felt every time she eased her boat out from the slip and left the city behind had kept her in New York despite losing her job.

Now, as she crossed the West Side Highway into the familiar territory of the parking lot leading up to the pier, Aldinger kept her focus straight ahead. She couldn't bear to look downtown at the smoke rolling from the towers. She hiked up a set of steep wooden stairs, unlocked the Manhattan Kayak Company office, and sent an e-mail message to her parents: *I'm okay. Details later.* Back downstairs, she was comforted by friends' familiar faces. Among them was John Krevey's wife, pier co-owner Angela Krevey. Aldinger was standing with her when her friend's face went white, her hand rising to cover her open mouth. It was 9:58 A.M. Aldinger spun around to see that all that remained of the 1,362-foot-high building that had been her workplace two months prior was a column of thick gray dust.

There was nothing Aldinger could do for her coworkers, but she could help the people around her. She opened up the kayak company office to passersby in need of working phones or water and then she joined in evacuation efforts on the pier. Three party boats would soon begin a free ferry service, circling back and forth to New Jersey. One of those boats was the VIP Yacht Cruises vessel *Royal Princess*, captained by Dennis Miano and crewed by the *Ventura*'s captain, Pat Harris.

After an unsuccessful attempt to fuel up his sailing yacht at a Morris Canal filling station and an aborted effort to rescue people from the seawall just north of the Battery with Jerry Grandinetti, Harris had decided to find a safe place for his existing passengers: the mate and his family. The timing of that decision led to what Harris singled out as "one of the great moments of the day."

Earlier that morning, as Harris explained it, the mother of *Ventura*'s mate Josh Hammitt had been on a shuttle bus headed to Newark Liberty International Airport when she'd spotted the burning towers. Worried about her family, who lived so close by in Gateway Plaza, she'd gotten off the bus and hitchhiked her way back east, winding up on the waterfront of Liberty State Park. She had been looking out over the Hudson at smoke and debris blowing across her apartment building, worrying that her children might be caught in the middle of it, when suddenly the *Ventura* pulled into Morris Canal with her family on board. "She just dropped to her knees," recalled Harris. "You could see a mother's relief flooding out of her."

Harris had just finished making fast the *Ventura*'s docking lines in a marina there when a man pulled up alongside the boat in a hard-bottom inflatable dinghy. He was diabetic, the man explained. Did they have any sugar? Welcomed aboard, the man hoisted himself over the side and asked to stay. Harris agreed and in turn requested permission to use the man's boat to cross back to Manhattan.

"I got a first aid kit, a bunch of towels, threw them in the dinghy," Harris explained. "I had to go to the other side to see what I could do." Wearing his NYPD safety vest and Harbor Unit Auxiliary hat, Harris set out for North Cove. "Halfway across the Hudson in this little tiny rubber boat I got this feeling like, *This is a very vulnerable position*." Before him the notch of waterfront that he called home lay dust-cloaked and swirling with smoke and debris. Did he really want to go back? *Am I going to do any good over there?* he wondered. *What's the use of this little first aid kit and these hand towels in the midst of all that?*

By now the Hudson was choppy, the river swarming with vessels. All their wake action left Harris bobbing about, feeling "very, very small." For the first time since he'd watched a jet slice through the North Tower, Harris let his guard down. "I didn't have anybody around me to be responsible for," he explained. "I felt vulnerable and not like I had to be in command and take charge." *Do I really want to do this?* he wondered. "But of course there was never any doubt." As Harris saw it, the situation that morning was "every mariner's best case and worst case scenario. . . . It's a horrible situation. *And* you're able to help."

Then the helmsman on a small Coast Guard boat spotted him. The boat sped over on high alert. But when the crew spotted Harris's NYPD garb, the boat slowed and they waved him along. "That waving me on put me over the edge at full speed ahead," Harris explained. "There was a little transfer from one mind to another that: We're on the same team. Let's all go in there and do something!" He steered the dinghy straight to the *Ventura*'s regular North Cove berth, a few floating docks away from where Grandinetti had pulled away in the *Excalibur*, which was now ferrying passengers and supplies back and forth around the harbor.

When Harris stepped onto the wooden dock, his feet sank several inches into the gray-white powder that coated every surface, every leaf on every tree, "like a heavy snowfall." Above him, long reams of flaming computer paper glowing orange corkscrewed about the air.

Heading up the ramp that led to the plaza, he spotted "somebody military" with a rifle. By nightfall, 750 National Guard troops would be stationed in the city. By morning that number would climb to 3,500. Then Harris noticed VIP Yacht Cruises owner Bob Haywood Sr. standing by the *Royal Princess*. "Can you help us get this boat out of here?" Hayward asked Harris. "I remember looking around and saying there's nothing I can do here, but I can certainly help there." So he stepped aboard.

Just then, recalled Harris, two bewildered tourists appeared on the scene, one German, the other from Ecuador. "They were kind of stunned, wandering around totally out of place and completely disoriented." They too boarded the *Royal Princess*, joining Harris and Captain Dennis Miano at the helm. The four of them became the ad hoc crew of the 125-foot, 200-passenger dinner yacht. Harris's first challenge was to excavate the ropes securing the boat to the pier. "I had to dig down past my elbows just to find the docklines, getting all that dust in my face, in my lungs, coughing and kind of digging through."

Captain Miano backed the boat from the slip and steered north. Finished with his decking duties, Harris made for the bridge, but to do so he had to cross through blizzard conditions on the open upper deck. The grayish-white powder lofted by the head wind engulfed him. "I remember saying to myself, *This is bad shit. You can't breathe this.*" But of course to continue aiding in the evacuation there was no avoiding it. Like hundreds of thousands of others, Harris took in lungfuls of toxic power, the true cost of which would only become apparent years later. In the moment, although he had some sense of the potential danger, Harris knew there was only one thing to do: keep moving forward.

When Harris reached the wheelhouse, he was covered in so much dust he was scarcely recognizable. Miano, meanwhile, had heard over the marine radio that people were queued up at Pier 63. We're going to take them across to New Jersey, he told Harris. Harris didn't think twice about being thrust into the evacuation effort. Participation was "instinctive," he explained. "It was right place, right time . . . It was just the right thing to do." But he had no intention of breathing in more of that dust than he had to. Down in the dining room, where tiny blocks of safety glass from a broken exterior door or window lay sprinkled across the dark, patterned carpet, Harris grabbed a white tablecloth off one of the neatly arrayed tables and tied it across his face.

It must have been about noon when the *Royal Princess* pulled up to Pier 63, recalled Harris. As Miano maneuvered the boat into docking position Harris readied the lines and took in the scene before him. He spotted John Krevey "doing crowd control" at his "little emporium." There must have been thousands of people there, recalled Harris. Krevey "had everybody queued up in a long snake like you would have in the velvet ropes. . . . He had everybody organized and flowing onto the boats. Somehow he put that all together." The crowd watched their approach. "We pulled up port-side-to," Harris recounted. "That's the side that had the broken glass. There was still all this dust all around you by the bulwarks and in the corners, and I probably looked like a snowman, as somebody called me later."

A voice called up from the crowd: Where the hell has this boat been?

From the wheelhouse Miano responded. "This boat was at Ground Zero."

It was the first time Harris heard the designation that later would become so iconic, so inextricably linked with the World Trade Center site.

Captain Miano gave Harris a clicker, instructing him to stop at 300 passengers, though the boat was only certified for 200. "We were going to be safe *and* bend the law," Harris explained years later. "He's a pro, I'm a pro. We both know that stability is built on a number of factors and we could push that limit and get these people out of there."

He was struck by how people's patience and the organized queues created a sense of calm in the midst of chaos. "Everything was beautifully ordered and when the boat was full, they just said okay and we pulled out," Harris recalled. Despite being a captain himself, Harris had stepped easily into a support role. "I was just along for the ride and part of the team," he said. The captain "gives the orders and you just say, 'Yes sir.' It worked out very, very nicely."

The German tourist had taken to his ferry deckhand role as well, employing what Harris called "good German efficiency." The gangway—more of a three-foot-wide plank with no handrails—extended from a lower deck at midship to the pier and Harris worked the boat end while the tourist stood by the end on the pier. "I remember him moving one arm in a circle and the other pointing just like a policeman would," recalled Harris. "And then on the Weehawken side we did the same thing," emptying the boat of passengers, level by level.

The *Royal Princess* looped into rotation with the Spirit Cruises and Circle Line boats dropping off in Lincoln Harbor. With each run, the organization on the New Jersey side escalated visibly. Among the boats "there seemed to be a natural pattern evolving," Harris said, ". . . a counterclockwise motion of flow outbound to New Jersey." Captain Miano ran the boat at a regular cruising speed of about six knots and, by Harris's recollection, the round trip took about 15 minutes. Though plenty of people boarding were dusty and disheveled, nobody was injured and just one woman, crying hysterically, was distraught enough to require extra care. Mostly, they were incredibly grateful, Harris recalled.

"There was a proud moment, and I often get choked up when people ask me or I think about it," said Harris.

> "That moment when we're disembarking people, I would say at least a third paused for a moment at the top of that gangplank and put their hand on my forearm and said, 'Thank you. Thank you for helping us out.' And I remember looking at their faces and seeing jaw lines that were set. Eyes that looked determined. And I thought, *This is a culture that we can be proud of.* There was cooperation. There was no panic. There was a realization that we'd been hit hard, knocked down, got back up, and we were determined to get through this. And I thought, *This is New York. This is America.* It was basically at the worst of times that we rose to the occasion. The horror of people's parts and the debris—that has less of an impression on me than the positive. I was seeing the paradox in humanity—the best of our kind at one of our darkest moments. That will choke me up more than the horror."

That day the *Royal Princess* made five or six round trips between Pier 63 and Weehawken, transporting as many as 1,800 people off Manhattan Island.

Just before the first tower came down, Pamela Hepburn had helped to evacuate children from a school in Lower Manhattan where she'd been working as a polling-place volunteer during the morning's election. After taking shelter in the school during the collapse, she'd slogged her way to the waterfront through the ankle-deep layer of gray powder and mangled paper blanketing the streets to borrow a bicycle that she figured would be the fastest way to reach her nine-year-old daughter, Alice, who was in middle school on Twenty-first Street between First and Second Avenues. Once Hepburn had collected Alice, she'd focused on finding somewhere to be, somewhere safe, where the air was clean. Their loft apartment on Murray Street, four blocks north of the trade center site, was out of the question. Who knew what condition it was in now? A tugboat captain who'd been working in New York harbor for more than two decades, she made a beeline for Pier 63. "That's where we hung out so much," she explained years later. "My little compass needle went right there."

At the pier she encountered friends, including Angela Krevey and several other mothers who were trying to busy their kids. Alice joined the other children, and Hepburn set out to help with the evacuation. The lines at Pier 63 were growing. Although her 1907 tugboat was currently out of service and under restoration, Hepburn also owned a whaleboat called *Baleen*—a 26-foot-long, Navy-issue, open motorboat traditionally used as a lifeboat, in rescue operations, or to transport personnel between vessels or to shore. *Baleen*, currently berthed downtown at Pier 25, was what Hepburn had available, so she biked south, secured the boat and a longtime acquaintance to help crew it, then returned north.

Back at Pier 63, she discovered that the big charter and dinner-cruise boats now making loops across the Hudson were all delivering passengers to Weehawken. "I thought, geez, there have got to be people who live in Jersey City," Hepburn recalled. She decided to evacuate people farther south. She spotted Bruce Rosenkrantz and asked him to find people who needed passage to the Jersey City–Hoboken City line. She could take 14 or 15 passengers, but warned that they might get a little wet.

At the foot of Twenty-third Street, two lines now serpentined around the parking lot. One line was for passengers boarding boats from Chelsea Piers. The other was for people boarding from Pier 63 Maritime. Among the thousands waiting were Chris Reetz and Chris Ryan, the sales associates from L90 on Twenty-third Street, now caught up in the general evacuation of the island and looking for a way home. Ryan scanned the crowd for familiar faces. And then it began to sink in. A lot of people had just died. He started going through his mental Rolodex thinking, *Who do I know that was down there?*

A man with a megaphone—John Krevey—was calling out directions, instructing people about which line to join. People continued to swarm to the waterfront, yet the lines remained orderly, with no shoving or cutting. As each boat arrived, crews loaded up passengers to capacity, and beyond.

Reetz and Ryan stepped into the Pier 63 line for boats heading directly across the Hudson to Weehawken, planning to hoof it to Hoboken once they reached the Jersey side. The line moved faster than they expected, but still they waited for close to two hours before reaching the front. Then someone—Bruce Rosenkrantz—yelled out, "Does anyone want to go to Hoboken?"

"We do!" they replied, thankful for the promise of a direct route home.

"I would have gotten on an inflatable boat," explained Reetz years later. "I would have gotten into a canoe or a kayak just to get off that island." And Pamela Hepburn exuded confidence. Reetz and Ryan were among the passengers who boarded the whaleboat for that first run. As soon as they boarded, Reetz let out a sigh. *I'm on a boat. I'm safe. These people are getting me to Hoboken.* He'd had no hesitation about climbing into this small boat rather than one of the larger passenger vessels.

Ryan, too, was struck by Hepburn's commanding presence, which was reinforced by the declaration by her acquaintance, now crew member, who proclaimed to passengers, "Pamela is one of the best tugboat captains in the harbor. She's been running tugs for more than 25 years." *She's been doing her job as long as I've been alive*, thought Ryan.

Baleen was weighed down, sitting low in the water—"gunwales to," as Hepburn expressed it—when she steered it out into the choppy Hudson. As the boat headed south, water splashed up over the sides. No one on board said a word. To this day Ryan recalls that Hepburn "was clear when all of us were fuzzy. . . . She was bringing me to safety and she made me feel safe while doing it regardless of all the crazy shit that was going on."

When Reetz saw the sunny, clear skies over New Jersey he felt a wave of relief. He was off the island, out of harm's way. It occurred to him that this was his first time on the Hudson River, and he dipped his left hand into the cool, salty water. When he looked up he saw downtown Manhattan engulfed in smoke. He thought of the destruction, the chaos, the death. The juxtaposition of the two skies—the blue sky on the right and the black sky on the left—overwhelmed him. *It's a shame*, he thought, *that my first time on the Hudson is on a beautiful day—like this.*

Hepburn had wanted to transport her passengers farther south along the Jersey waterfront, but docking *Baleen* posed specific challenges. Her goal was to serve the most people by making multiple quick trips across the river, but she needed a tie-up site that could accommodate a boat that rode so low in the water. Through the years she'd spent working in the harbor, Hepburn had been trained to look at the shoreline as a resource to help deal with all sorts of issues: "loss of power, beaching the boat, or putting a line out to get beer." In this case she needed a spot where passengers could disembark safely. Local knowledge pointed her to Long Slip, an old, narrow pier with railroad tracks located at the Jersey City–Hoboken line. She didn't expect to be greeted by police when she arrived.

Long Slip was by no means an official drop-off point. As Hepburn steered her whaleboat toward the pier, Ryan, who had worked on boats when he was younger, could tell that this would not be the easiest landing. The pier was nothing more than a raised slab of concrete with rocks jutting out at the waterline, and Hepburn nosed in with no fendering to cushion the impact. Ryan heard a crunch as the boat made contact and it occurred to him that this captain was risking damage to her boat in order to ferry them home.

In an instant, uniformed police pounced. They ordered the boat away, saying that no one could get off. Hepburn didn't flinch. "Excuse me," she said firmly. "We've got people on here who've just walked to Twenty-third Street from the World Trade Center. They need to disembark." That was enough. Her crewmate was already wrapping the bowline around a spike that stuck out of the concrete.

"We were like friggin' refugees," recalled Ryan. "And she was a leader." When the police backed off and began helping, Ryan was struck by how they had succumbed to her authority. "She did that with her confidence and her clarity." Instead of blocking their path the police helped people out of the boat and up the rocky ledge.

Next the passengers faced decontamination. The Hoboken Fire Department hazmat team, under the direction of the Hoboken health officer Frank Sasso, who was concerned about the potential of a chemical or biological attack, had established a makeshift decontamination facility, building a giant shower out of PVC piping and in-line spray-heads fed by three fire engines. More than 10,000 people were decontaminated that day.

Nearby St. Mary's Hospital had set up a field hospital unit where more than 2,000 people were ultimately triaged. Unlike at the triage center just north of the Colgate Dock, most of those arriving in Hoboken were uninjured. According to EMS reports, only 179 patients required hospital transport, and most did not need medical care.

Reetz recalled being surveyed for injuries, sent through a sprinkler system of sorts, then doused with water by a firefighter cracking open a fire hose. Reetz might have wound up soaking wet, but at least he was safe in New Jersey, not far from his apartment. Now his head swam with a mix of relief and disbelief. *What the hell just happened? What do we do now?*

All across the country, people were asking those very same questions, but not all with the same level of urgency as those caught up directly in the aftermath of destruction, including those still working in New York harbor.

PART THREE

THE AFTERMATH

> I dream'd in a dream, I saw a city invincible to the
> Attacks of the whole of the rest of the earth;
> I dream'd that was the new City of Friends
> —Walt Whitman

CHAPTER 10

"We *have to tell us what to do.*"

ALL MORNING NEW YORK WATERWAY'S PORT CAPTAIN, Michael McPhillips, had been fielding questions. NYPD Harbor Unit officers, Port Authority representatives, and captains from a handful of different tug companies, among others, contacted McPhillips for information and sometimes direction. Usually his job involved tracking vessel schedules, managing captains and deckhands, vessel maintenance, and Coast Guard compliance. Being thrust into this position of responsibility out of scale with his position overwhelmed him. When a tug captain called asking where he should pick up and drop off passengers, McPhillips mistakenly sent him to a pier with nowhere near enough water to accommodate such a deep-draft vessel. "I caught it before they got in there, but it would have ripped out the bottom of their boat," he recalled. "It sucked, honestly. I had all these other boats calling that were coming in to help and I had to make all these decisions."

Given McPhillips's prominence on the radio, it made sense that when Coast Guard leadership arrived on scene shortly after the second tower fell, an officer contacted him by VHF. "He asked what was going on. I explained it to him," McPhillips said. "I asked if I could go over capacity. He responded with a 'yes'. Then he asked if we needed help. And I said, 'Absolutely.'" A short while later, McPhillips heard the VHF radio broadcast calling for "all available boats" to aid in the evacuation.

Though neither can say for certain, the Coast Guard officer who called McPhillips that morning may well have been Lieutenant Michael Day aboard the Sandy Hook Pilot boat *New York*. After the boat had dropped marine inspectors at strategic points along the Battery to help guide evacuees toward boats and prevent vessel overloading, it continued its long, slow "barrier patrol" along the tip of Manhattan, swooping around from the Hudson to the East River and back again.

The goal was visibility. Not only did this sweep afford the personnel aboard the pilot boat good vantage points for looking up both sides of Manhattan to watch for points of traffic congestion or other issues, it also made the boat itself, with the U.S. Coast Guard ensign waving, more conspicuous. Day sought to empower mariners with the idea that the Coast Guard was on scene, present, and available.

From his post in the wheelhouse as the boat moved up the East River, Day watched a ferry pull off Pier 11 carrying a boatload of passengers. A great cheer erupted as the boat backed away. Day found it strange to see people so happy to leave Manhattan, though he only needed to shift his gaze slightly to see the smoke and be reminded why.

Despite the efforts of the inspectors, Day saw vessels precariously loading far more passengers than they were designed to carry. Some mariners even radioed to *tell* him that they were carrying more people than permitted. They weren't so much asking for permission as reporting the fact, Day recalled. While witnessing botched Haitian and Cuban migrations on overloaded boats earlier in his career, Day had seen the horrors of drowning refugees clawing for capsized vessels. Now he felt torn. "We were trying to get as many people off the island as we could," he said. But, he acknowledged, "If a boat flipped over we'd have even more people in the water because of my actions. And I was responsible. I felt responsible."

Like McPhillips, who couldn't reach his supervisors, Day was making on-the-fly decisions that normally would have been beyond his authority. When mariners requested permission to violate regulations, Day explained, "I rogered, laughing at myself a little bit. It was just like, *wow! I broke more rules than probably I've enforced in my whole Coast Guard career.*" He later joked that at the time he'd reassured himself with the thought, *I'm just a lieutenant. What are they gonna do to me?* In truth, making leadership calls normally left to those at "flag level" weighed on him.

But Day's rule breaking was far from reckless. Like many other mariners who took a more flexible approach to compliance that morning, Day made considered decisions in response to unprecedented circumstances. As disaster researchers James Kendra and Tricia Wachtendorf point out in their book *American Dunkirk: The Waterborne Evacuation of Manhattan on 9/11*, "Just because rules were broken does not mean that there was a lack of order, organization, or concern for safety." Instead, rules were being:

> "thoughtfully disregarded, even in the desperation of those first hours when people just wanted to do *anything*. We call this *rule breaking with vigilance*. Everyone broke the rules, but they broke them gracefully, with sensitivity for consequences and with a sure-footed sweep through a potential minefield of possible mistakes and accidents."

Crucially, the violations still reflected the guiding principle behind the rules—"Taking positive action to make things better," as Day described it. And the infractions arose in direct response to extraordinary conditions. "It got easier to break them as time went on," Day explained. "I won't say it didn't matter . . . But you know what? They're looking at a burning hole. It didn't really matter in the balance of it." Ultimately, Day did his best to encourage captains to act safely while continuing to summon additional boats in hopes that more vessels on the scene might reduce the impulse to overload.

Part of what gave Day the confidence to make on-the-fly judgment calls was the leadership style espoused by his commanders, Captain Harris and Admiral Bennis, which emphasized encouraging their people to take the initiative to make their own decisions. "I really felt when I worked for Admiral Bennis that I was totally empowered to do the right thing," explained Day. "Do the right thing and I'll take care of you. Don't worry about it. I mean as long as you can say this is why and the reason."

"I'm a huge believer in empowerment," Bennis affirmed in an interview with a Coast Guard historian months after the attacks. His longstanding leadership style had been to encourage his team to "go out and make magic and be brilliant." His approach on September

11, once he finally made it back to New York at about 3 o'clock that afternoon, was no different. "What I did is what I always do. I went in with the folks. I sat down with them. I got a briefing—the first of thousands of briefings. I asked very few questions—some pointed questions just to be sure we're going on the right track," Bennis explained. "I had a team that I had complete trust in, and I let them know that right up front.... I just tried to stay there in the midst of them, but, absolutely, I never micromanaged them."

In fact, that approach had actually trickled down to Bennis and Harris themselves from the highest ranks of the Coast Guard. From his earliest communications with his superiors following the attacks, Bennis was reassured that he and his team would be trusted to do what needed to be done. "They were all pretty confident with our abilities and capabilities," he explained of his higher-ups, who were located all across the Northeast.

> "I knew what I wanted to do. I knew from working with the city the best way to accomplish it.... But I wanted to know, was I in fact a free agent? And I was. As the commandant put it out later, he said he allowed his field commanders to let their creative juices flow and do what they needed to do, and I was able to do that."

Instead of establishing a top-down command and control structure, the Coast Guard, from the top brass down to the on-scene rank and file, allowed for the organic, needs-driven, decentralized response that played an enormous role in the ultimate success of the boat lift. This approach, in turn, allowed mariners to take direct action, applying their workaday skills to these singular circumstances, without being stifled by red tape.

As Kendra and Wachtendorf explain: "Even with an eye for security and safety, [Coast Guard officials] were still able to recognize the value of an improvised citizen response to the terrorist attack." Instead of interfering with the waterborne evacuation that was already under way, the Coast Guard *participated*. Commanding officers, both on- and off-scene, granted their blessing, legitimizing the spontaneous, unplanned evacuation through facilitation and support, thereby encouraging more mariners to get involved.

No one had foreseen the sudden need for evacuating a huge swath of Manhattan Island. Yet as terrorized people continued to flee to the waterfront, more and more boats turned up to rescue them. To Harris the white wakes visible in aerial views over Lower Manhattan looked "like the spokes of a wagon wheel." Mariners were already responding. "Nobody had to be told to help," Harris explained. "The asking, basically all that did was put people on the right frequency. People were already primed."

As greater numbers of vessels and evacuees amassed along the shoreline, streamlining operations became the biggest challenge. By midmorning, so many mariners had joined in the effort that the regular passenger piers jammed up with boat traffic, thwarting the vessels most suited for using those piers from efficient operations. "The only way to fix it was to get organized," said Harris. That organization was implemented in large part by Day and the pilots operating aboard the *New York*, which continued its barrier patrol. Their efforts were made easier by the relationships that both the Coast Guard and the Sandy Hook Pilots had with the New York harbor community.

Day's initial broadcasts from the helm of the pilot boat set the tone for the Coast Guard's position of cooperation and participation rather than interference with or controlling the efforts already under way. "United States Coast Guard aboard the pilot boat *New York*," Day began. "All mariners, we appreciate your assistance." Rather than ordering people around, he and most of his Coast Guard and pilot colleagues did their best to leverage their existing relationships with members of the New York harbor community to foster a team approach.

"The New York maritime public probably responds better to someone they know than someone they don't," Day explained. "New York as a city and the maritime community in particular, is built on relationships." Day's history of "externally focused" work within the Coast Guard, including the year he'd spent in an "industry training" exchange program, helped him have more engagement with the community than a typical Coast Guard officer.

In 1998, he had worked with the Port Authority in 1 World Trade Center, meeting with maritime industry people to learn about the impacts that Coast Guard regulations and actions had on commerce. He had also been working with the Harbor Operations Committee, which held regular meetings bringing the Coast Guard together with commercial operators, the Sandy Hook Pilots, the Port Authority, the Army Corps of Engineers, and other harbor stakeholders to "seek nonregulatory solutions" to port problems. "I was in a unique position to understand relationships between the Coast Guard and the public," Day explained. "As a result of it I had a degree of trust."

What struck Day that morning, and stuck with him thereafter, was what he called the "clarity of purpose: hey, we're doing a good thing to help people." Helping others "is a core ethos of the maritime community," he explained. "It's just part of the culture. . . . You're at sea and someone needs your help and you'll divert hundreds of miles out of your way to help someone."

Day also recognized that the Coast Guard's regulatory functions and role as "enforcers" could end up as a divisive force if not carefully managed. Day was mindful about fostering "a unity of effort," as the guiding principle of operations that day. One choice that helped serve that approach was the decision to join forces with the Sandy Hook Pilots and use their boat as a floating command center. By nature of being a law enforcement and regulatory agency, the Coast Guard would, of course, have some clashes with boaters during a normal day. The pilots' daily operations, however, routinely included more exclusively collaborative relationships with other harbor operators.

Harris explained the success of the evacuation's collaborative approach this way:

> [Day] "had a really good relationship with the maritime industry in the port. But it wasn't just him. Also on that boat was Andrew McGovern.

Andrew was recognized, had a lot of personal leadership power. Between the two of them, it became conversations. Nobody demanded anything. Nobody yelled at anybody. Nobody ordered people to do things. Everybody said what they wanted to do and the guys on the *New York*, the pilot vessel, made it possible for them to do it. They could talk to them. They knew them. They had sat through meetings with them for years.

Aboard the *Chelsea Screamer*, Captain Sean Kennedy didn't count heads. Instead, he and crewmate Greg Freitas focused on loading as many people onto the 56-passenger thrill-ride speedboat as they could, as fast as they could. "We filled it up. If we peaked it, it was by only a few." And then the captain shot straight across the Hudson to the closest New Jersey dock: at Liberty Landing in Morris Canal. As passengers disembarked, Kennedy took a few empty water bottles and filled them from a hose on the pier. He wanted to be able to offer water for people to wash their hands and faces, to clear their mouths and throats.

On his second run, Kennedy headed toward a cluster of people in Battery Park who walked clutching clothing to their faces. They'd gathered four blocks south of the World Trade Center near Pier A, a historic municipal pier built in 1886 that had stood vacant since 1992. Kennedy called out to a firefighter on land there, asking him to cut the lock on the gate that prevented people from reaching the water's edge. Seeing the gate opened, people scrambled toward the boat.

After Kennedy had offloaded more passengers back in New Jersey, the camera crew that had originally chartered the boat as a platform for shooting footage earlier that morning told him they needed to submit tape to their office near Rockefeller Center. Could he run them back to Manhattan anywhere near there? Kennedy said he could drop them off at a pier near West Forty-sixth Street. Hearing this, some passengers asked if they, too, could disembark there instead of New Jersey.

As he set up to land on a barge at the south side of the *Intrepid*, an aircraft carrier that's now part of a sea, air, and space museum, Kennedy saw mobs of people queued up to board Circle Line and World Yacht boats from a nearby pier. He spotted a man waving and trying to get his attention and sent his crewmate Freitas to run over and ask what the man wanted. The man was looking for a way to cut the line. He offered $4,000 cash if the *Screamer* would deliver him and three others across the river now. But Freitas refused. "That's how desperate people were to leave immediately," Kennedy explained. "Money didn't matter."

Already off the island, Karen Lacey had no money. And no shoes. When she stepped onto the abandoned pier in Jersey City, the Merrill Lynch director was numb, but not with cold. Although her clothes were soaked through with Hudson River water, here on the Jersey side the sun shone bright, warm, and unobscured. "I was thinking about shoes," recalled Lacey, "about having to take a fairly long walk without them." Her Hoboken apartment stood about two miles to the north, but Lacey planned to make a stop along the way. "We'll stop at Modell's and get sneakers," she told Tammy Wiggs. Although Lacey, having finally dropped her bag in the river, was without a wallet, she couldn't imagine the store clerks, seeing her wet and gray, would turn her away. "They'll give us shoes just to get to Hoboken and I'll come back and pay for them later," she figured.

As the two women walked through the streets, Wiggs barefoot and Lacey in shredded stockings, people couldn't help but notice them. For the most part they were "gracious, not gawking," Lacey remembered, though she did overhear a few whispers: 'Oh my God. They were *down* there.'

Dozens of passersby offered their help: 'Do you need water? I have shoes. I can give them to you. We'll go upstairs. We'll get it for you. We'll be right down.' But, spurred on by the promise of a shower, Wiggs and Lacey refused the overtures and kept walking.

Twenty minutes into their journey, they reached Modell's only to find it shuttered behind metal security screens. "That's when it hit me," Lacey explained. *It's 10:30 in the morning; Modell's isn't open.* Consumed by shock, panic, and the instinct to flee, her mind had circumscribed the morning's events as some "uniquely New York thing." She'd somehow imagined that once she got across the river "people were going to be having lunch and selling shoes." Realizing that the store had sent home its employees on a Tuesday morning somehow cemented for Lacey the gravity of the World Trade Center attacks. There was nothing to do but continue on.

The superintendent of her building on Hudson Street in Hoboken unlocked the door to her sixth floor apartment. As she pulled together some clothes for her young colleague to borrow, Lacey asked herself which would be the least offensive, cleanest looking pair of panties she could loan, then chose a pair of purple Calvin Kleins.

"I never thought I'd wear someone else's underwear," said Wiggs.

"I never thought I would offer them," Lacey replied. "But they're there if you want 'em."

A black ring marked the tub after they each finished showering. No matter how many times Lacey blew her nose, what came out was gray.

Wiggs felt like she had "little glass shards" in her eyes. She had no lens solution available, yet was so desperate to wear her contacts so she could see that she swished the lenses around in her mouth in an effort to clean them.

When Lacey's family came by the apartment, Wiggs hid in a back room. She declined Lacey's offer to join them when they went out to eat at a local restaurant. Instead she used Lacey's landline to call everyone she knew, and finally connected with a friend who offered to drive her to her parents' home in Baltimore.

Also walking barefoot through Jersey City was Florence Fox, still carrying four-year-old Kitten, who peppered her nanny with questions. "Where are we going? When are we going to get there?" Fox enlisted her help with the search. "We have to look," she told her charge. "Kitten, can you see anything? Can you see something that looks like a hotel?" The little girl had grown calmer but Fox could tell she was still scared by the way that she clung. That fear would end up affecting the child for years to come. "I was talking to her just to make sure she was okay."

They eventually found a hotel. The lobby was mobbed with people trying to book rooms, but none of the others were covered in dust. Fox strode to the front of the line. "Maybe it was arrogant, but I really didn't care." "Can I get a room?" she asked. "They must have thought I was crazy." The clerk was sympathetic but couldn't help. Every single room was booked. Throughout the area, people left stranded by the closures of the region's three major airports were hauling their luggage through the streets—some hitchhiking, some pushing commandeered airport luggage carts—on the hunt for buses, rental cars, and hotel rooms.

Eyes burning, skin itching, and barefoot, Fox had carried Kate through the streets only to end up at a hotel with no vacancies. As she begged, then argued with the desk clerk to no avail, a few hotel guests, in town on business, offered to take them in. They showed Fox to a room, then left the two alone to get cleaned up.

Fox washed Kate first. The girl was shaking. Instead of drawing her a bath, Fox stood Kate in the tub, letting the shower water run so she wouldn't have to sit in the stew of toxins that rolled off her body. Once the little one was clean and wrapped in a towel, Fox prepared to bathe herself. "You can sit on the floor," she told Kate, then climbed in and closed the shower curtain. Kate screamed.

"Don't close the shower curtain!"

"I don't want you to get wet."

"I don't care if I get wet," the girl pleaded. Fox looked at the frightened child and saw that she was changed. *She's right*, Fox thought. *After what we've been through what is water?* So she showered with the curtain open.

Once they were both clean, Fox distracted Kate with cartoons and picked up the phone. She still couldn't retrieve her employer's phone number. Instead she dialed her sister whom Fox knew could reach the girl's mother.

Then the girl and her nanny curled up on the bed to wait. "I just remember feeling so scared. And feeling cold. And looking at Kitten and being so afraid for this child." The people who'd taken them in brought food, but Fox doesn't remember eating. She doesn't remember exactly what happened next.

Although Kitten remained separated from her parents for nearly 12 hours during the frightening ordeal of fleeing her home, her bond with Fox, who loved the girl like the two were family, offered her safe haven throughout. While finding trust and comfort among perfect strangers was one of the hallmarks of the disaster, the solace of familial ties could not be underestimated.

Shortly after the first tower fell, beginning law student Gina LaPlaca had aligned her fate with the men who'd saved her from stumbling down the subway station stairs. But now, temporarily blind, eyes bandaged in a Manhattan hospital, she decided she needed to be with family. When somebody finally managed to reach her mother on the phone, the frantic and relieved woman arranged to have LaPlaca's uncle, John Coyle, who worked near East Thirty-third Street, pick up her daughter in the hospital. The two men waited with LaPlaca until he arrived, and then it was time to say thank-yous, swap phone numbers, and issue heartfelt (yet ultimately unrealized) promises that they'd all keep in touch.

LaPlaca found great relief in being taken under her uncle's wing. Coyle, who had helped raise LaPlaca after the death of her father, signed her out of the hospital and explained the facts of the situation: "Your neighborhood's off limits. They're not letting anyone in there. They're evacuating everybody. You're gonna go home with me to Staten Island." With that he hooked his arm in hers and pointed them toward the Staten Island Ferry, three miles to the southwest.

But first they stopped at a street vendor to buy some large-framed "Jackie O" sunglasses. Not only was LaPlaca feeling self-conscious about her bandages, any hint of light that made it through the dressings irritated her injured eyes. For an hour the young woman was guided, unseeing, through her new city. She registered the shifts between neighborhoods by sounds and smells. Chinatown struck her as particularly loud and full of activity. Then, as the two proceeded farther downtown near the trade center, the streets went quiet. "You would hear sirens but there weren't really cars on the street, or the usual activities there," she recalled. Walking "as close to the water as we could," they crossed Wall Street, and soon enough LaPlaca arrived back where she'd started—where she'd wandered blindly through the dust cloud that morning. The smell caught in her throat. Now that she was in the care of her uncle, the weight of the day began to sink in.

The ferry terminal hummed with people. All morning James Parese and his fellow Staten Island Ferry captains had continued making runs back and forth to Manhattan, delivering more than 50,000 civilians off the island. On return trips from Staten Island, they transported emergency workers and supplies. Parese, who had started his workday at 5 A.M., wouldn't finish until 5 P.M.

Coyle weaved LaPlaca through the crowd to a spot to wait where they would be visible. In addition to a ferry, the two were waiting for LaPlaca's friend and houseguest who had undergone her own trials that day, wandering the Manhattan streets in ill-suited footwear and relying on the kindness of strangers. Through communicating with the friend's mother, LaPlaca's mother had coordinated the friends' reunion so that both young women would have a safe place to stay on Staten Island.

"There's a blond girl waving," said Coyle. "Is that her?"

The two friends embraced. "Thank God you're all right. What happened?"

"Oh my God. What happened to you?"

When they boarded the ferry, Coyle, "a creature of habit," shot straight for the same seats that he chose every day on his commute. All around LaPlaca heard the "whispered, small conversations" of shocked fellow passengers. Not until she was safe at the West Brighton, home of her uncle did LaPlaca notice her battered feet, all blistered and torn from her heeled sandals.

Wet as Chris Reetz, Chris Ryan, and Ryan's girlfriend were after the decontamination hose-down that greeted them in Hoboken, walking home made the most sense. But they all planned to meet up a little later. Reetz knew he needed some sort of support system—someone to rely on and someone with whom he could process all that had happened. "Chris," he explained, "was the closest thing I had to family."

McSwiggans Pub had become "a home away from home" for Ryan and his girlfriend. "We knew every bartender. We'd been to the owner's house. We knew his kids." So that was where the three went and waited for their fellow regulars to file in. "It was a weird kind of feeling in that bar," explained Ryan, an "uncomfortable happiness" about everyone who'd survived that revealed itself through nervous laughter. He recalled wanting to lighten the mood and then realizing that was not the right thing. "This was not a time to be funny. It's like telling a joke at a wake or something, and in many ways it was. I mean, holy shit. Everybody in there knew somebody who was dead or could have been dead. Or they didn't know yet if they were dead." And so the trio stayed late, sitting, and drinking, watching the television, and waiting for their friends to appear. Once they'd found this "stable place," they didn't want to go home.

For Rich Varela, home came as an afterthought. While Lacey reunited with family, Wiggs drummed up a way out of town, Fox secured a hotel room, LaPlaca headed for Staten Island, and Reetz and Ryan sought comfort in their neighborhood bar, Varela had remained at the post he'd been assigned by firefighters on the *John D. McKean*, which was actively pumping water to land-based engine companies. Varela doesn't recall what the *McKean* crew had asked him to do, exactly, but at the time he understood that his duty was to stay put. Before long it seemed like he was the only one left aboard. *What am I doing here?* he wondered, noting that while he stood wearing just a life vest everyone around him wore breathing apparatuses and protective gear. Still he didn't leave—not until a firefighter (maybe Tom Sullivan, maybe someone else) came back and asked, 'Why are you still here?' While the *McKean* crew would remain on station for days, the telecommunications specialist who'd volunteered his assistance was now free to go.

Relieved of duty, Varela immediately thought, *How the hell do I get out of here?* For the second time that day, he needed help getting off Manhattan Island. He walked north toward North Cove and stepped out atop the southern breakwater to scope out his options. Then a small, maybe 30-foot, aluminum boat pulled up, and a man called out over a loudspeaker.

"Hey, buddy. You all right? You need a lift?"

"Yeah."

"Where do you gotta go?"

"Jersey."

The guy nosed the boat in close enough that Varela could jump on.

Across the river, Varela stepped off the boat in the shadow of the Colgate Clock (the iconic 50-foot diameter octagonal clockface that had been overlooking the Hudson since 1924, when it stood atop the now-razed Colgate-Palmolive factory) and was greeted by triage center medics. They looked him over, gave him wet cloths to wipe himself off, and then directed him to the free buses that would deliver him to Newark Penn Station. Not until he settled into a seat on a New Jersey Transit Raritan line commuter railcar, at somewhere around three o'clock in the afternoon, did Varela process what he must look like wearing a life vest on the train. Neither the shock nor the generous purpose for which he'd sacrificed his shirt muted his embarrassment.

By late afternoon, when crowds amassing along the shoreline began to dwindle, tugs, dinner boats, ferries, and other vessels shifted duties from delivering passengers *off* the island to ferrying emergency workers and others *onto* Manhattan as well as transporting goods. Throughout the evacuation and well into that first night people had sought to cross the river toward Manhattan as well as away. Some felt duty-bound because of their professions—firefighters, steel workers, doctors, nurses, journalists, and canine rescue squadrons among them. Others were desperate to find loved ones.

The Lincoln Harbor Yacht Club's general manager, Gerard Rokosz, fielded pleas from a man worried about his pregnant wife, fireboat *John D. McKean* transported a father concerned for his son, and New York Waterway captains reported denying passage to reporters. Coast Guard Boatswain Carlos Perez and his crew shuttled police officers; Spirit Cruises Operations Director Greg Hanchrow ferried fire department personnel, police, and "a lot of suits"; and on Jerry Grandinetti's first run back to Manhattan from Owl's Head, Brooklyn, he delivered medical personnel to North Cove.

The rescue efforts, meanwhile, demanded supplies. And like the evacuation, Coast Guard Lieutenant Michael Day explained, the supply runs "just kind of happened." The highly visible, 185-foot pilot boat waving a Coast Guard ensign provided a ready hub for rescue workers' requests. When firefighters approached asking for drinking water, Day and his colleagues made calls to the New Jersey Office of Emergency Management. Civilians on the Jersey side cleared store shelves, piling bottles along the waterfront. When Day radioed requests to mariners to run the supplies across the river he was inundated with volunteers. Rescuers needed dust masks and eyewash, wrenches and diesel fuel, and acetylene for torches to cut steel. Day and his team made requests to New Jersey and mariners delivered whatever could be gathered to their aid.

It was dusk when Day first stepped off the pilot boat *New York* into the "eerie gray snow" of Ground Zero. The mixed crew of Coast Guard and Sandy Hook Pilots personnel had established a good working relationship with emergency officials in New Jersey, and now Day was walking around the site, hoping to locate someone from New York City's Office of Emergency Management. "There were body parts everywhere," he recalled. The sight of a foot still in its shoe transfixed him. He couldn't take his eyes away. *This is a war. A siege,* he thought, eyeing the National Guardsmen patrolling the streets of Manhattan with M-16s.

Structural fires continued to rage around the World Trade Center site, orange glows penetrating the darkness left by power outages. The air hung thick with debris, the hovering dust spotlighted in the beams of the few construction light boxes that had so far been erected, courtesy of Weeks Marine. When the wind kicked up it lofted charred papers, sending them fluttering through the air.

At last the message that nanny Florence Fox had sent through her sister reached Kitten's family. The four-year-old's mother was on her way. Silverton had been searching for her daughter since morning. As she sprinted from the townhouse to the nursery school and everywhere else she could think of to look, the towers came down around and on top of her, covering her with ashy gray soot. Only when firefighters warned her of a possible explosion and ordered her to board a boat did the distraught mother evacuate.

On the New Jersey side, Silverton ran into people she knew. She asked them if they'd seen Fox and Kitten. Little did she know she was closer now to her daughter than she'd been all day. Finally, at about eight o'clock that night, a call came through. Her daughter was safe, and not far away.

"I was expecting this almost fairytale reunion," recalled Silverton. "Kate would run into my arms and I would pick her up and swing her around. And then hug Florence. We'd all hug. And instead I got off the elevator and they were coming down the hall. Florence was carrying Kate. I'm not sure if Kate even recognized me." This was the first sign that Silverton, a psychologist, had of the post-traumatic stress disorder that would afflict the child in the days to come.

"We just cried and cried," recalled Fox.

By three o'clock in the morning, Jerry Grandinetti had wound up less than 500 feet from where he'd started on the *Excalibur* that morning. All afternoon and evening Grandinetti had worked delivering evacuees off the island, transporting personnel around the harbor, and carrying supplies into North Cove. Many of the requests he'd answered came from the pilot boat, which was now tied up along the outside of the breakwater at the south end of

North Cove. Grandinetti wound up tying the *Excalibur* to the inside of that same breakwater. Still aboard were the two NYPD Harbor Unit officers who'd been assigned to help him. But not for long. In the predawn hours, a sergeant boarded.

"How are you guys doing?" Grandinetti recalled the sergeant asking the officers.

"We're good to go till daylight, Sarge."

"No you're not," their superior replied. "You're done."

Relieved of their posts, the officers left, and Grandinetti was suddenly alone with his thoughts. "I didn't know what to do then. I cracked a beer and sat in the wheelhouse. Then I decided, maybe I'd try to get upstairs to my apartment to close the windows and get my wife some clothes." He knew the two of them wouldn't be allowed home for a long while and this might be his best chance to collect some essentials.

With the power out, Grandinetti had no choice but to climb the stairs to the nineteenth floor by flashlight. Before he'd fled that morning he'd been hanging out a window with his camera. That window was still open when he returned, and now every object, every surface in his apartment was plastered with dust. He didn't stay long before heading back downstairs and retreating to the boat—the best refuge he now had.

Grandinetti sat in the wheelhouse. *What do we do now?* Despite his exhaustion he couldn't settle his mind. *I can't sleep*, he thought. *I can't sleep looking at this red glow.*

What the fuck just happened? From the helm of the go-fast boat that he had commandeered that morning, Spirit Cruises Operations Director Greg Hanchrow tried to wrap his head around the events of the day.

All afternoon the Spirit Cruises dinner boats had ferried passengers from Pier 61 to Lincoln Harbor. Hanchrow and the *Spirit of New York* crew had to adjust their makeshift gangway system as the tide shifted, but still they managed to offload passengers safely and comfortably, down a ramp and through a main-deck cargo door. But by seven or eight o'clock that night all three Spirit boats were docked back at Pier 61. What had started that morning as a mission to rescue company assets by pulling vessels out of Manhattan had turned into a rescue mission for people, and ended with the boats secured back in their berths. By now it was getting dark, and Hanchrow had decided he'd better bring back the boat that he'd appropriated to Petersen's Boat Yard & Marina in Upper Nyack. Only a few lights shone at the trade center site where smoke continued rolling into the night sky.

Heading north up the Hudson River, his head swirling, Hanchrow looked over to his right at the West Side Highway. Suddenly it hit him. There were no cars. Instead a caravan of about 10 huge quarry earth-moving trucks with enormous tires lumbered south. *This is 12 hours later and we're already moving material*, he thought.

His next thought was: *I'm gonna have to figure out how to get down here tomorrow.*

When at last Hanchrow arrived home, emotion overwhelmed him. "I was sad, crying. I was mad. I'm very upset." He thought of his daughter. *Here I am in my nice cozy home with my daughter we've had for three months who's still speaking Bulgarian. Well, I'm glad I brought her to this country. Maybe she would have been better off over there.* With "a fucking big glass of rum" in hand, Hanchrow decided to push off some of the hard questions until later. *All these things are gonna have to be addressed at some point. But not now,* he thought. *There's shit that's gotta get done.* He might not yet know exactly *what* needed doing, but he was determined to make his way back onto Manhattan Island the next morning to figure it out.

In hindsight, Hanchrow called his plans to return "pragmatic." *I'm not in any danger,* he figured. *No one's shooting at me. There's nothing imminent.* Instead he felt a need "to go and be centered around the boats, around the pier, in the city saying, 'Okay, what are we gonna do?'" Despite all the day's uncertainties, Hanchrow was absolutely sure of one thing: "We need a time-out, and we need a huddle. . . . No one is going to tell us what to do so *we* have to tell us what to do."

Chapter 11

"Sell first, repent later."

AT DAYBREAK ON WEDNESDAY, SEPTEMBER 12, Sean Kennedy once again stood at the helm of his thrill ride tour boat *Chelsea Screamer*. With him were crew member Greg Freitas and more than a dozen National Guardsmen who had slept aboard another of Kennedy's boats, a dinner yacht called *Mariner III*. Kennedy had happened across the para-rescue specialists at about eleven o'clock the night before while they'd been preparing to set up camp outside at Chelsea Piers. When he'd offered better accommodations, they'd accepted, some settling into available cabins while others stretched out on deck, "still in their armor, guns by their sides."

Now it was dawn and, as promised, Kennedy was transporting them back to their duties. Smoke continued to roil up from the Pile—all that was left of the collapsed buildings—as he tied the boat at North Cove. The National Guardsmen climbed ashore and then set off between two buildings. Kennedy never saw them again. Just south of the cove, he spotted a friend, Huntley Gill of retired New York City fireboat *John J. Harvey*. Kennedy shuffled across the plaza to greet him, his feet disappearing beneath the ashy powder that reminded him of new fallen snow.

All night the fireboat had continued pumping, supplying river water to hose lines stretched inland. Chief Engineer Tim Ivory had barely slept in the hammock he'd strung up on the bow of the boat to ensure he could be called upon at a moment's notice. Several times he woke in a panic, fearing an air raid when U.S. fighter jets swooped low overhead. Gill, too, had slept fitfully, if at all, up in the wheelhouse. Now, as he and Kennedy spoke, Gill reached into the dust at his feet. "Look at that," he said, pulling out a card printed with the name "Windows on the World"—the famous restaurant that had occupied the top floors of the North Tower. Although little had yet been confirmed about the numbers and identities of the people who had been killed, it was already understood that, in all likelihood, the people in the restaurant when American Airlines Flight 11 hit had perished.

Having dropped off the National Guardsmen, Kennedy and Freitas decided to knock on doors to offer transport to anyone still in the area. Ultimately they evacuated about 20 people, many with pets, delivering them to Chelsea Piers.

Located directly up the Westside Highway from Lower Manhattan, Chelsea Piers facilities had served a major role in disaster response from the earliest hours of the attacks. By 11 A.M. on the eleventh, EMS triage centers that had been disrupted at the trade center site were relocated to sound stages at Chelsea Piers. A 160-bed trauma hospital was established in the studio space of the television show *Law & Order*, and ambulances from all over the region marshaled outside, awaiting dispatch. The piers offered water stations for people fleeing from downtown and became a key debarkation point for the waterborne evacuation, delivering more than 10,000 people off the island from its docks.

Then, the following day, more than 30,000 people arrived to volunteer their help and connect with other New Yorkers. Families of the missing gathered at the Chelsea Piers Field House and met with counselors. In the days that followed, thousands of uniformed personnel were fed in an events center at Pier 60, hundreds of rescuers slept and showered in Chelsea Piers facilities, and truckloads of donations and supplies were assembled and processed for delivery to the trade center site.

At North Cove, donations were stacking up. Supply runs had continued throughout the night, and the esplanade was quickly overrun. Just as they had for centuries, before containerization pushed shipping off the island toward Brooklyn and New Jersey, Manhattan's shores had become littered with break-bulk cargo— a wide variety of goods that had passed hand-to-hand onto vessels before being offloaded, also by hand, onto the dusty moonscape surrounding what would soon become known universally as Ground Zero.

"The supplies were coming in so fast and furious that we were running out of room and we couldn't distribute them fast enough," recalled Coast Guard Lieutenant Michael Day. By Wednesday, the flood of private citizens' donations had been supplemented with massive corporate contributions arriving in bulk. Mariners and land-based volunteers moved pallets of Levis, cat and dog food, galoshes, and flashlights—all manner of goods, some more useful than others. So many boats were pulling in to offload that the radios became useless for managing traffic in, out, and around North Cove. Instead a Coast Guard crew member assigned to the bow of the pilot boat used hand signals to direct vessel movements. "It was a hands-on VTS," explained Day. All this activity prompted discussions about where else boats might land supplies. As Day explained, the knowledge and experience of pilots and local tug captains helped immensely. Chelsea Piers and the Battery were identified and formally established as two additional supply depots.

Stacks of boxes and garbage bags full of donations covered more and more of the waterfront plaza around North Cove. John Doswell, the Pier 63 regular who had helped with crowd control during the evacuation, had hitched a ride aboard a small boat to fireboat *John J. Harvey*, where he was also a crew member, to offer what help he could. He and his wife, Jean Preece, wound up using a fire hose to clear an expanse of the plaza's hexagonal paving stones of the thick layer of dust that kicked up in the slightest breeze. Now at least donations could be kept slightly cleaner, and offloading would be easier.

Although Greg Hanchrow couldn't have predicted exactly what September 12 would require of him, he'd felt certain that a huddle was the best way to find out. So, first thing on Wednesday morning, at a coffee shop named Ruthie's just north of Pier 61, Hanchrow met with a handful of Spirit boat captains and other mariner friends to brainstorm.

"It's gotta be a disaster down there," said Hanchrow. "There's no place to stay, no place to sit. There's no water, no toilets, no phone lines." And what if there were more bombs? Wouldn't it be wise to station a "big evacuation device" nearby? Hanchrow called upon his old friend, Gene O'Hara, retired from the NYPD Harbor Unit, in hopes that he might help. "Is there any way we can get the *Spirit of New York* into North Cove?" he asked.

In less than 24 hours, the situation had changed entirely. While all manner of vessels had moved in and out of North Cove on the eleventh, helping with the evacuation, the FBI quickly took the lead in controlling the World Trade Center site and deciding who could come and go. "That's a tall order," replied O'Hara. But he offered to call his former colleague, Harbor Unit pilot Tony Sirvent.

Another mariner, Vince Graceffo, who had worked the deck aboard the *Spirit of New York* during the evacuation, chimed in to piggyback on Hanchrow's idea: What about food? Maybe they could feed the rescue workers, too? He said he knew a chef at the Tribeca Grill. Located in the "frozen zone" that had been shut down by authorities, the restaurant would be closed for the foreseeable future. (Soon officials would erect 12,000 feet of six-foot-high chain-link fence that would bisect the neighborhood called TriBeCa, for Triangle Below Canal, sealing off a zone patrolled by National Guardsmen wearing battle fatigues and helmets and carrying automatic weapons.) Maybe the managers would consider donating their fresh food that would otherwise go to waste. Then there was the well-stocked Spirit Cruises commissary. With the harbor closed and the country reeling, Spirit wouldn't be running lunch or dinner cruises anytime soon. The Spirit executive chef figured they had enough food on hand to last at least a few days. *We could do this,* thought Hanchrow. They just had to secure permission.

A short time later, Sirvent nosed the bow of a police launch against the floating docks on the eastern edge of North Cove so that Hanchrow and O'Hara could jump off. Advised by Sirvent that there was a command post at the corner of Vesey and West Streets, Hanchrow and O'Hara trekked inland through the eerie landscape toward the heart of the disaster. As they walked past the green-domed 1 World Financial Center building, Hanchrow looked up at the trees and noticed a pair of pants dangling. *That's a fucking strange place for a pair of pants*, he thought. He surveyed the damage to the building whose whole first floor was blown apart and piled high with rubble. Looking up at the tree a second time, Hanchrow spotted what he hadn't registered at first: sticking out of the bottom of the pants were two feet. The sight was more than he could grasp.

The two kept walking, but geography was obscured by wreckage. Hanchrow kept shaking his head. *Unbelieveable. Unbelieveable.* "I didn't even know which direction to go." Finally they found a police chief who said approval for nonemergency services assets would have to come through the Jacob K. Javits Convention Center. The two would head there next. But first, they knew they'd best also secure permission to move the boat through the harbor.

Back on the waterfront, the Coast Guard flag flying atop the pilot boat *New York* offered some hope of answers. "All right, you're the cop guy, I'm the Coast Guard guy," Hanchrow told O'Hara. "Let's go see if the Coast Guard's there."

Hanchrow recognized pilot Andrew McGovern from the Harbor Ops meetings he'd attended, so he approached him first. McGovern steered Hanchrow to a Coast Guard officer. "Listen, we have the *Spirit of New York*, a dinner boat, and we'd like to bring it into North Cove and provide relief service here," petitioned Hanchrow. "We can feed 600 to 1,000 people an hour. The boat is geared and designed to have high turnover of people, and we don't have anything else to do."

The officer didn't say yes, but he did convey urgency, explaining that if Hanchrow had any intention of making this happen he'd better do so within the next 24 hours. As Hanchrow recalled it, the officer told him: "This is going to turn into something that you've never seen. Nothing's gonna come or go." Lower Manhattan was, after all, the site of a massive act of terrorism, and now formal structures for dealing with the aftermath were being pulled into place by multiple local and federal agencies. With that, Hanchrow and O'Hara headed back to the police launch for transport back to Chelsea Piers to work on setting the plan in motion.

Given all the transportation shutdowns, the easiest way for Hanchrow and O'Hara to get from West Twenty-third Street to the Javits Center at West Thirty-fourth was to walk. But they hit a roadblock at Thirty-third; the street was blocked by a tank and piles of sandbags. Blue-shirted cops were everywhere. *This is fucking crazy*, thought Hanchrow. The surreality didn't subside in the convention hall itself. Already, on the day after, thousands of people had gathered to offer up their particular skills and resources.

"I start getting in the line where the big boys are with the big toys," recounted Hanchrow. "There's this guy in front of me who's talking to all these white-shirt, suit people and he goes, 'I own an ironworking company and I got three trucks and 60 guys and they're parked in a parking lot outside the Lincoln Tunnel. You say the word and I've got 60 guys and welding machines and they're coming down the West Side Highway.' 'Okay, go talk to this guy,' they told him."

Next it was Hanchrow's turn to make his case. As he described his idea for a floating relief center that would offer shelter and food, as well as a mobile command post that could double as an evacuation vehicle, "their ears perked up." All right, you go talk to this guy, they told him. It sounded promising. But the next people he spoke with told him they'd get back to him and left it at that.

What do you mean you'll get back to me? thought Hanchrow. *What is this, a job interview?* More than "a little perplexed," he walked back down to Chelsea Piers.

Convinced that this was too good an idea not to execute, Hanchrow and O'Hara decided to move ahead regardless. O'Hara offered to call his girlfriend who used to know someone who worked for the governor. Hanchrow, meanwhile, called his boss. When the acting president for Spirit Cruises, who was working out of the Boston division, heard Hanchrow's plan for a four-day relief mission from Thursday through Sunday he replied that he'd watched the New York boats evacuating people on CNN. "Do whatever you need to do," Hanchrow recalled him saying. "You have my permission to use the assets any way you see fit."

The plan was taking shape. Spirit's chef loaded up food on the *Spirit of New York*. The boat crews loaded fuel from the *Spirit of New Jersey* to the *New York* in a completely unorthodox overwater transfer "with a big fucking Rube Goldberg kind of hose." Hanchrow scheduled Spirit crews to work round the clock. The *Spirit of New Jersey* would become a warehouse for donated supplies. "All during this time I'm like, *Of course this is gonna work*," Hanchrow said. "*It's a big boat. We've got a lot of food. We've got a lot of people. There's going to be a lot of need for all this down there.*" His guiding principle had come from his father, who'd always said: "Sell first, repent later."

While Hanchrow and his team focused on the logistics of feeding rescuers, and Lieutenant Day and others helped usher supplies onto the site, the repercussions of shutting down the largest container port on North America's East Coast loomed large. Although the physical infrastructure of the Port of New York and New Jersey hadn't been directly affected beyond the destruction of the Port Authority headquarters in the North Tower, the whole harbor had been shut down, unleashing a ripple effect of serious commercial consequences.

Once he'd collected himself, after crawling out of the rubble of Tower Two and seeking refuge in the Coast Guard building at the Battery, the Port Authority's port commerce director, Richard Larrabee, had set to work reassembling his team and trying to get the port reopened. Not until he was safely inside, watching the North Tower fall on television, did the retired admiral fully grasp how narrowly he'd escaped being swallowed up by wreckage. He showered, changed into the spare Coast Guard jumpsuit that someone had given him, and then called his wife (who hadn't yet known his whereabouts) to tell her he was alive, but didn't know when he'd be home.

He caught a ride aboard a New Jersey police boat to the Port Newark waterfront and walked inland to a Port Authority administration building. Personnel there had suffered a clear, cross-harbor view of the unfolding disaster. "They were all just completely beside themselves," Larrabee recalled.

His first job was to track down his team. Remarkably, almost every one of the 100 people under Larrabee's direct supervision had made it out of 1 World Trade Center safely. (One man who had been attending a meeting in another part of the building did not survive.) Next, Larrabee would leverage his longstanding relationship with Rear Admiral Richard Bennis (at one time Larrabee had been his boss) to advocate for opening the port as swiftly and safely as possible.

The port commerce director spent the night at the Port Authority building, then early on September 12, he greeted one of his staffers, Bethann Rooney, who'd just returned by train from Washington, D.C., where she'd been speaking about homeland security, "before homeland security was the buzzword," as she explained years later. Rooney's boss hugged her, saying, "Thank God you're okay." Then he announced: "You're in charge of security."

Thus began her 14-year effort, first as security manager and later as general manager of port security, to create a security program "from nothing." "Security didn't exist in the maritime industry on 9/11," explained Rooney.

> "In terms of the Port Authority it was about crime prevention and loss prevention for cargo, and public safety-type matters. This was the same in every port around the country. Your piers, your docks were wide open. So on September 10, 2001, you could drive down to the port, take out your lawn chair, and sit down and have your sandwich and sunbathe at lunchtime right adjacent to the side of a ship unloading cars from Japan, completely open and accessible."

Big changes would be needed, both nationally and internationally. But here in New York harbor, the Port Authority and Coast Guard immediately aligned forces to find answers to some daunting questions. Later that morning, in a conference room in Port Newark, Admiral Bennis, Lieutenant Michael Day, numerous Port Authority police officers, Rooney, Larrabee, and others assembled to discuss what Rooney referred to as the new paradigm. "What is this thing called maritime security? The port is closed. When are we going to reopen it? How are we going to reopen it? What does that look like when we reopen it?"

On everyone's mind was the threat of more terrorist activity, recalled Larrabee. "The decision to close it was easy," he said. More difficult was "getting people to agree on the process by which you could reopen the port."

The Coast Guard was confronted with two opposing missions: protecting potential targets of a second-wave attack and continuing commerce in a port that regularly handled approximately 6,000 inbound and outbound containers daily, the closure of which created complications worldwide. The port shutdown prevented the movement of essential commodities that were crucial for running infrastructure—oil, gasoline, and aviation fuel—along with perishable goods like fresh fruit. Then (as now) the Port of New York and New Jersey was the nation's largest port for refined petroleum products. "Within 20 hours I was getting calls from the White House saying they're running out of gasoline in Portland, Maine," recalled Admiral Bennis. "New York is a huge petroleum port. Huge. We supply the petroleum to Logan Airport and all these other places, and people don't realize that."

Costs ran high, and both port and Coast Guard officials were keenly aware of the potential for longstanding economic damage resulting from the closure. "We operate in a very, very competitive world. If you ship goods from China," explained Larrabee, "you are looking for reliable, time-sensitive, low-cost factors to move your goods. You could send your goods through Los Angeles or Long Beach. You could send them through Norfolk or down through Halifax, or you could send them through Newark. The longer the port was closed, the less competitive we were, and the more likely it was that cargo could be getting shifted in other directions." Just as Commander Daniel Ronan had feared, the port shutdown would indeed cost $1 billion each day. The pressure was on.

Additional pressure stemmed from the transportation disruptions, some of which would continue over the long term, as well as the intense need for onloading and offloading supplies to and from the trade center site. The limitations of existing waterfront facilities were immediately apparent to people like dock builder Paul Amico whose expertise lay in creating bridges between water and land.

On the afternoon of the eleventh, even in the midst of helping to coordinate safe docking options and load passengers, Amico was already considering what infrastructure would be needed on the twelfth. He managed to get a message out to his shop to start building the hinge plates that he knew would be necessary to connect gangways to the shore. He knew he had a number of gangways in his yard, left over from use during OpSail 2000, and was sure he could find key places along the waterfront where they could be put to good use. "I knew we were going to have to put temporary facilities up to get supplies in, and ferry terminals to get civilians in and out."

One crucial first step toward building new waterfront access points was to identify conditions along the shoreline. This meant knowing what was underwater, out of sight at the water's edge. New York Waterway immediately made arrangements for a diver as well as a boat with a side scanner to map out, on September 12, the riverbottom terrain between the Holland Tunnel Ventilator at Spring Street and the northern edge of Battery Park City. They discovered that the area was littered with pilings that remained from the eight or so finger piers once existing there.

"A diver would come up and say, 'There's no way we could get a boat in there,' or, 'Yeah, we can get a boat in here,'" Amico recalled. This underwater reconnaissance would prove critical when, by Thursday evening, DonJon Marine began plucking moorings out of the riverbed to make way for Weeks Marine to come in with cranes and barges to begin loading out World Trade Center debris.

But before all that, at dusk on the night of the eleventh, Amico and the upper management of New York Waterway met at Pier 11 on the East River with representatives from New York City's Office of Emergency Management to discuss waterborne transport options. As Amico recalled, the basic message from the office to Waterway was: *Do what you guys do.* That meant something to the effect of: figure out how to move people from Point A to Point B over water. The questions that naturally followed included: *What docking facilities can we pull together? Where can we put them?* and *What boats are available?* Another meeting convened shortly thereafter at Waterway's headquarters in Weehawken to drill down to the specifics. "What we wanted to do was take a look at piers, pier heights, pier conditions, and what boats are available," recalled Amico. Waterway management jumped on the phone to call anyone who owned a fishing boat that might be used to shuttle passengers. And so the creation of an ad hoc system of ferries, terminals, and routes began.

From September 12 on, with subway routes changed and the downtown PATH train tracks buried beneath World Trade Center rubble, ferries became a key mode of transportation. To compensate for disrupted downtown service, the PATH ran a free cross-river service uptown between Thirty-third Street and Newark and Hoboken, New Jersey. On September 13, the Port Authority reopened the Lincoln Tunnel, George Washington Bridge, and Port Authority Bus Terminal. But most subway stations south of Fourteenth Street remained closed.

City streets south of Canal Street weren't reopened to regular traffic until October 10 (and then only during certain hours). But businesses in other parts of the city began operating within days after the attacks. Indeed, some never closed. This being New York, city life and the commuting routines of many quickly resumed. To meet increased demand, New York Waterway added seven new routes. The company chartered vessels from up and down the Northeast coast, including whale-watching boats from Massachusetts, in order to run as many as 35 ferries during rush hour. Still the ferry system was stretched to capacity, and Port Authority officials announced that they were racing to build a new ferry terminal near Battery Park to cut New Jersey commute times in half.

On an average weekday before September 11, 2001, 32,000 passengers had ridden New York Waterway ferries. By December, that number jumped to nearly 60,000. "If September 11 had not happened, we would have reached this level of ridership in five to six years, but we reached it almost overnight," company president Arthur Imperatore Jr. told *The New York Times* at the close of 2001. "It has raised the awareness of a lot of new people about the capability of ferries. They're not a toy. They're serious transportation." To a large extent, he argued, the ferries helped hold Lower Manhattan together for months following the attacks.

The Staten Island Ferry had also done its part to normalize commuting routines when service resumed on September 17, despite new regulations prohibiting vehicles aboard the boats. By late December, the New York City DOT reported that total ridership aboard both public and private ferries had more than doubled since the attacks. By April 2002, ferry use grew 91 percent, reaching the highest levels since the 1940s. All told, nearly a dozen new ferry routes had been established, and over the course of six weeks, an existing Battery Park terminal was reopened, and a railroad barge was retrofitted to create a new six-slip ferry landing at Pier A.

Amico was upgrading the New Jersey side as well. On the second weekend following the attacks, he was working to revamp a small pier float at Colgate, just north of Paulus Hook and directly across the Hudson from North Cove, which previously could only serve a single boat. To accommodate the variety of different watercraft now arriving on the scene, each with different docking configurations, he had to adapt the pier with both bow- and side-loading capabilities. The job involved driving pilings, and he took his usual approach, checking to see if the pilings that he was installing were vertical by lining them up with a tall building. "I turned around to see if the pilings were straight by looking at the World Trade Center, like I would usually do," he recalled in an interview some months later. "Of course, the towers were gone. And I burst out in tears."

―――――⚭―――――

As new infrastructure was built to accommodate the "new normal," the crucial work of defending the harbor continued. Taking on that responsibility were Coast Guard personnel like Boatswain's Mate Carlos Perez, regularly assigned to New York harbor, as well as Coast Guard members (reserve and otherwise) who'd been called in to serve the region. With the establishment of an incident command center at Fort Wadsworth, 1,500 active duty, reserve, and auxiliary personnel from Coast Guard stations nationwide were called in to report for round-the-clock duty. Twenty-six additional Coast Guard cutters and small boats were deployed to New York harbor. "We turned into a true response organization for about six weeks," explained Acting Commander Patrick Harris.

It was around 9 P.M. on Tuesday when Perez steered the utility boat, *41497*, toward the dock at Coast Guard headquarters for the first time since the initial search and rescue alarm had sounded that morning, more than 12 hours earlier. For all that time he had been patrolling the harbor, enforcing the port closure to all nonessential vessels, and evacuating civilians. Finally back at Fort Wadsworth, he was greeted by a sea of new faces waiting at the docks to service the boats as they arrived. The whole station was mobbed with people, and "yet it was very quiet"—so quiet that Perez could hear boots stomping as they made their way across the grounds. After a three-hour break for food and rest, he and his crew returned to the water, working straight through until 8 A.M.

When they returned the next morning, even more personnel had amassed. More than 200 people from all across the country stood mustered, recalled Perez, awaiting the arrival of the captain of the port. The announcement "Attention on deck!" called the crowd to order, and the reverberation of so many feet coming together to stand at attention "brought a chill down my spine." For Perez, the echo embodied camaraderie, reminding everyone that, whatever their differences, they all stood side by side in support of a greater good.

CHAPTER 12

"Thanks for your help!"

ON SEPTEMBER 13 AT 8 A.M. THE COAST GUARD REOPENED the Port of New York and New Jersey, with significant restrictions in place. Commercial and vessel movements required Coast Guard approval. Passenger vessels carrying more than 50 people had to have uniformed security personnel aboard. Recreational traffic was permitted only in certain areas during certain hours. Vessels were subject to Coast Guard boarding and inspection. Meanwhile large vessels (weighing more than 300 gross tons) were subject to further restrictions, including mandated advanced notice of arrival; provision of certified crew lists, including nationality; and other constraints. Anchorages remained closed, and commercial port traffic in general was significantly constricted under these new rules.

Similar issues confronted land-based traffic as the Port Authority reopened area bridges and tunnels, as well as the bus terminal on that Thursday. Two days of bridge, tunnel, and road closures into Manhattan led to widespread disruption of commercial deliveries of all sorts.

Fortunately, the food stockpiled aboard the *Spirit of New York* successfully reached its destination in North Cove. At dawn on Thursday morning, Greg Hanchrow stood at the helm, nosing the 35-foot-beam, four-deck dinner boat toward the 75-foot-wide opening into North Cove. Although nothing had yet come of their Javits Center visit, and no official permission had been granted, the Spirit team had decided to set out anyway. Perhaps because of Hanchrow's discussions with the Coast Guard aboard the pilot boat the previous day, no one made any efforts to stop them. For his part, Hanchrow was less worried about being stopped by authorities than he was—suddenly, upon his approach to the cove—by the sheer physics of what he was attempting to do.

"This feeling of dread came over me when I looked at the size of the opening," explained this seasoned captain who'd confronted countless sticky docking situations in his decades at the helm. "I was like, *I don't know if this fucking thing is gonna fit in here.*" Then, moments later, after he'd gunned the boat through the opening, he wondered if he could turn the dinner yacht around. With a bit of maneuvering, Hanchrow was finally able to round it up and back it in, sidling the boat's port side against the eastern edge of the southern breakwater. Although he did knock the mast off one sailboat, and the bow did stick out into the opening a fair amount, the boat was in, and now the crew focused on making fast lines and setting up gangways. VIP Yacht Cruises loaned Hanchrow a giant set of folding stairs to help keep the *Spirit of New York* accessible in North Cove through the tide swings.

Once the vessel was secure, a new round of doubt settled in. "Everybody on the boat was like, what are we gonna do? How is this gonna work?" Hanchrow didn't have answers. But O'Hara chimed in with reassurance. They'd let a few rescue workers know the boat was here and the news would spread by word of mouth, he said. "Within literally an hour, or an hour and a half, there were mountains of people coming into North Cove," recalled Hanchrow. The food the Spirit chefs had thought would last them until Sunday was gone in eight hours. "By the end of the dinner shift on Thursday there was nothing left on the boat."

Hanchrow figured they'd need a small craft to shuttle workers and supplies to fuel their operation, so he contacted New York Waterway's Michael McPhillips. At first McPhillips said he could arrange a ferry to run back and forth every four hours. But soon it became clear that, with all the transportation disruptions, every available Waterway vessel was needed to ferry commuters. Instead, *Chelsea Screamer* Captain Sean Kennedy volunteered to deliver raw ingredients and kitchen crews to Ground Zero. City Harvest trucks (which normally distributed foods from restaurants, farmers, and grocers to soup kitchens, food pantries, and other city food programs) arrived at Chelsea Piers filled with food for delivery south to North Cove by boat. So many donations arrived that staff had to clear additional space on the *Spirit of New Jersey*, which was serving as a mini warehouse, holding supplies until they could be distributed.

All over Manhattan, word spread among restaurant owners and chefs that food brought to Chelsea Piers could be transported south to the trade center site. For the next two weeks the *Spirit of New York* operated as Chefs with Spirit, run by some of the city's top chefs and restaurateurs including Don Pintabona, Danny Meyer, David Bouley, Daniel Boulud, Gray Kunz, Bobby Flay, Tom Colicchio, Charlie Palmer, and others. Working around-the-clock shifts, the dinner boat's galley turned out as many as 600 meals an hour.

About a week into the Chefs with Spirit operation in North Cove, Hanchrow's bosses at Spirit Cruises told him that the Red Cross had contracted the company to keep the boat on station for another 14 days. "That was the only form of permission we got," said Hanchrow. "Prior to that no one ever asked any questions about how or why we were there, other than where the food was coming from."

In the early days that followed the attacks, Hanchrow and the Spirit team weren't the only ones devising ways to serve the needs of rescue workers. Along the perimeter of what was now known as "The Pile," volunteers were busy establishing outposts to provide food, shelter, supplies, healthcare, and compassion. In St. Joseph's Chapel, on the ground floor of Gateway Plaza, volunteers created a warehouse of boxes, doing their best to divide the mountains of goods into organized stacks. Soon the chapel became known as "St. Joe's Supply."

The small but efficient operation was being coordinated by Perry Flicker—a New Jersey office worker who wore his nickname, Flick, on a strip of masking tape across his orange vest—one of the small band of volunteers who had remained on site to help after the formal evacuation. A number of these volunteers were camping out in the offices of VIP Yacht Cruises, with permission from company owner Mark Phillips, who also granted them use of his golf cart for transportation. Flick's goal was to create a system organized enough that a rescue worker could call out a need for boots, or hydrogen peroxide, or water, and have it instantly provided. St. Joe's Supply was just one part of the ad hoc village that began sprouting up from the ash and debris around Battery Park City within the first 24 hours after the attacks. Elsewhere, makeshift outdoor bistros provided hot meals and space to rest, along with medical, massage, and chiropractic care.

On his repeat trips back and forth to the trade center site in the days that followed September 11, *Ventura* sailboat Captain Patrick Harris watched as the area that had once been his home evolved into a functional disaster recovery site. Once the evacuee lines at Pier 63 dwindled, Harris had stepped off the *Royal Princess,* grateful for the chance to spend the night with family in Weehawken. His own boat was safely docked in Jersey City. But that didn't stop him from returning to North Cove time and time again over the coming days to offer whatever assistance he could. When he heard that firefighters refusing to leave the site were surviving on quick naps on park benches, Harris brought over deck cushions. When cases upon cases of bottled water amassed on the shoreline, he took his place in the line of people passing the boxes hand to hand.

What struck Harris in the aftermath was the "cooperation, the spirit, the overwhelming sense that we were stronger than the bad guys." One moment still stands out for him as an embodiment of the collaborative ethos of those early days. At a picnic table erected in the cul-de-sac at the corner of South End Avenue and Liberty Street, Harris watched as the owner of a local liquor store—known both for its good wine selection and its "grouchy" proprietor—scooped heaping plates full of macaroni and cheese from steaming trays. With each plate he handed out, the owner included a paper dust mask.

"You know, 'Praise the Lord and Pass the Ammunition,'" explained Harris, citing the patriotic song that was written as a response to the attack on Pearl Harbor. The anthem recounts the story of a chaplain who, when asked to pray for seamen shooting at enemy planes, puts down his Bible, mans a gun turret, and fires, shouting the song's title line.

Other mariners who continued finding ways to offer their services after the boat lift included Greg Freitas. He'd worked with *Chelsea Screamer* Captain Sean Kennedy shuttling supplies and personnel until Friday when he felt compelled toward a larger purpose. Rain was falling on Friday morning so Kennedy decided to take down the white event tent from the deck of his other charter vessel, *Mariner III*, and bring it down to North Cove to protect all the supplies from getting ruined by weather. When they arrived, Freitas recalled, "We found about 120 volunteers with nothing to do. After asking, 'Who is in charge here?' and getting no reply, I said, 'Okay, I am in charge.'"

Freitas and the volunteers set up the tent, sorted and laid out the donated goods, then dubbed the operation The General Store. "By the next evening, I had acquired four more tents, a perimeter fence, tables, my own private police force, three golf carts, 200,000 baby wipes, and a red badge with the highest clearance from the mayor's Office of Emergency Management," recalled Freitas, adding that from Friday until he left at 8 P.M. on Tuesday, September 18, he managed a total of just five and a half hours of sleep.

The supplies that Freitas and the other volunteers distributed to an estimated 4,000 to 5,000 rescue workers included: respirators, blankets, candy bars, ear plugs, eye wash, safety goggles, gloves, sweatshirts, socks, shovels, underwear, toothbrushes and toothpaste, pick axes, Band-Aids, pants, hard hats, climbing ropes, flashlights and batteries, dog food, and dog-paw mittens. Many donations were accompanied by notes or children's drawings. "On one envelope there was a self-portrait in pencil of Nicole with the scrawled caption 'Thanks for your help!!'" Freitas recalled. Below it, an adult had added: "From Nicole, Age 6, New Jersey. She was so proud to put $2.12 in the envelope."

A large package from a pharmacy in Old Greenwich, Connecticut, arrived with a note that read: "7:45 A.M. Friday morning—It's raining—Here's some umbrellas. I started crying when I tried to stuff them in a bag. Couldn't stop. I finally started to feel like this happened to me too, like I'm a part of this f---ing thing too."

On that rainy Friday morning, other mariners had already begun gearing up to provide another essential service: debris removal. Emergency dredging crews with Weeks Marine worked round the clock to remove nearly 200,000 cubic yards of mud and silt to gain sufficient depth for setting up cranes and equipment. Within weeks, the company would design, mobilize, and begin operating barge-loading facilities in Lower Manhattan to transport what was estimated to be 1.2 million tons of wreckage piled at Ground Zero.

The ability to carry 150 truckloads worth of material in a single barge propelled by a single tugboat would play a huge role in the city's recovery efforts. Moving the rubble by water would not only prevent massive clogs in already congested tunnels and roadways and spare the city from having toxic dust roll through the streets, it would significantly reduce clean-up time. By May 2002, Pier 25, the main transfer area, would wind up handling a total of 1.18 million tons of debris, having shipped 62,581 barge loads off Manhattan by boat.

In the immediate aftermath and beyond, boats and boat crews provided essential assistance in smaller, simpler ways as well. The crew of *Powhatan,* DonJon Marine's oceangoing tug which was tied up near North Cove, for example, offered rescue workers the use of about a dozen on-board bathrooms. And in some cases marshaling supplies entailed not transport but the reallocation of on-site resources. When firefighters said they needed diesel to run land pumpers, Reinauer's Ken Peterson and others ran down the row of tugs along the seawall near North Cove filling five-gallon pails from the boats' stores. (After having arrived on the island at about 11:30 A.M. on Tuesday, Peterson stayed, working 24-hour shifts, until Friday night.) And when vessels began running low on fuel, the crews on well-supplied boats offered replenishment from their own tanks.

The camaraderie between mariners played a significant role in the success of both the rescue and recovery efforts. "Everybody worked very well together because New York is such a small harbor that a lot of us know each other," Peterson explained. One of the ways mariners from various sectors of the port knew each other was through the Harbor Ops Committee.

Many participants scheduled to attend the Harbor Ops meeting (that, as luck would have it, had been rescheduled to occur at the Coast Guard building at the Battery at 10 A.M. instead of at 9 A.M. in the World Trade Center because a conference room was not available) ended up coordinating and facilitating the boat lift. This was no accident. The individual connections forged though this committee, comprising more than 60 firms, organizations, and individuals—all stakeholders in New York harbor's multiuse waterways—helped the boat lift's smooth facilitation.

Rooted in a multiperspective understanding of the complex dynamics of the port, the committee worked to address, among other issues, the specific challenges posed by a mixed-use harbor where recreational traffic, including jet skis and kayaks, shared the same constrained waterways with commercial traffic like freighters and cruise ships. The committee's job was to "successfully anticipate problems, apply creativity, evaluate strategies, and develop solutions," as the Port Authority's then-port commerce director Lillian Borrone (Admiral Richard Larrabee's predecessor) explained in early 2001. Although she hadn't foreseen in March the events of September, Borrone's assessment of the working partnership illustrated perfectly the preexisting, collegial relationships that made the boat lift so successful.

As many New York harbor mariners tell it, the maritime community's collaborative spirit, and the overriding compulsion to help those in need that is so deeply entrenched it's encoded in maritime law, often eclipses other concerns, including competition between rival companies or potential friction with a law enforcement body such as the NYPD Harbor Unit for example. Pilot Tony Sirvent explained it this way: "That's what I try to tell these guys without a maritime background. These are licensed captains. The boating world is a community. We have a different relationship with them than most cops in the street have with other people." It's a relationship that is "absolutely" far less antagonistic and more cooperative.

For his part, Harbor Unit rookie Bill Chartier was struck by how seamlessly the boats worked together to get the evacuation job done. After he and his crew had yanked off railings in South Cove, said Chartier, "the ferryboats fell in line with us in just a continuous, amazing [effort]. It wasn't coordinated at all but everybody worked very calmly and very efficiently." By day's end, the Harbor Unit boats had evacuated as many as 5,000 civilians to New Jersey and Staten Island.

Coast Guard Boatswain's Mate Carlos Perez attributed the evacuation's success to the mariners' ability to keep focused on the main objective: "Making sure that we got people to safety helped us keep calm and stay the course," he said. That selfless concern for others was demonstrated over and over again that day, and in the days that followed. While in the immediate aftermath first responders were bracing themselves for another explosion or another building collapse the real danger, however silent, would turn out to be altogether more lethal.

Chapter 13

"They'd do it again tomorrow."

IN THE DAYS FOLLOWING THE BOAT LIFT, New York Waterway Port Captain Michael McPhillips was in charge of waterborne transportation for the New York National Guard, and transporting family members of the deceased to the trade center site. As such, he was among the many first responders who spent weeks—some for up to 18 hours a day—breathing the toxic air.

When the planes hit, the explosion of 91,000 liters of jet fuel started fires that then ignited an estimated 100,000 tons of organic debris, 490,000 liters of transformer oil, 380,000 liters of heating and diesel oil, and fuel from several thousand cars in the underground parking garage.

When the towers fell, the smoke from these fires combined with the destruction of the buildings to produce a plume of toxins and irritants containing a complex mixture of chemicals—including the combustion products of jet fuel, soot, metals, volatile organic compounds, and hydrochloric acid—as well as particulate matter from pulverized building materials and contents—including cement dust, glass fibers, asbestos, gypsum, heavy metals such as lead, polychlorinated biphenyls (PCBs), benzene, and dioxins. Among these were numerous well-known carcinogens. Researchers estimate that between 60,000 and 90,000 first responders were exposed.

Two years after his time in the dust cloud, McPhillips became too ill to work. The mariner who had run away to sea at age 16 wound up developing pulmonary issues and end-stage liver disease that he said resulted from dust exposure, and he was forced to retire from the industry. "I eat all the right foods," he explained. "I never drank—I never drank in my entire life." Still, he said, he's on the liver transplant list. "I have stage 4, grade 3 cirrhosis. And it's grade 4, stage 4 when they do the transplant."

Today he serves as director of social services and benefits for the FealGood Foundation—named after its founder John Feal, who was injured while volunteering at Ground Zero—that provides ailing first responders and victims with medical and financial assistance. As part of this work, McPhillips has spoken before government officials more than 100 times, advocating for the James L. Zadroga 9/11 Health and Compensation Act. "We're a casualty of war," he explained. Among those lost, he counts three New York Waterway captains who he said died from September 11-related health issues.

It's impossible to know exactly how many of the mariners who participated in the evacuation wound up suffering from illnesses related to their service. Even counting the overall number of mariners involved in the boat lift itself poses challenges. A total of more than 150 boats—vessels of all shapes, sizes, and purposes—were tallied by maritime consultant and event producer John Doswell. But, as evidenced by its exclusion of the Staten Island Ferry's boats, for example, that list is certainly not comprehensive.

Complicating matters further, each of these vessels could have been operated by between one and fifteen or more crew members. Reason would suggest that approximately 800 (and possibly many more) mariners actually worked on the water, participating in the evacuation that day, while still others provided essential assistance from shore. Reinforcing this estimate is the list of approximately 800 recipients of the Department of Transportation's 9/11 Maritime Medal, awarded in a ceremony in the rebuilt Winter Garden on September 17, 2005, "for meritorious service and sacrifice in the wake of the horrific terrorist attacks."

Calculating how many of those have since suffered from September 11-related conditions is equally difficult. A search for terms like *ferry captain, deckhand,* and *mate* made by data crunchers at the World Trade Center Health Program (WTCHP) counted at least 120 mariners currently registered with the program, 53 percent of whom are suffering from at least one illness or condition that doctors and researchers say is related to World Trade Center exposures.

The Zadroga Act, signed into law in early 2011, created the WTCHP to provide monitoring and treatment services through a Responder Program to rescue and recovery workers (including nearly 17,000 New York City firefighters), as well as treatment through a Survivor Program to those who lived, worked, or attended school in Lower Manhattan at that time. To date, the list of covered conditions—illnesses directly linked to trade center site exposures—includes more than 90 health conditions and more than 60 types of cancer. Demonstrating a clear connection between September 11 exposures with the various diseases that followed has taken years of research, based largely on studies tracking the higher rates of illness among exposed people than the general population.

Fifteen years after the attacks, the Centers for Disease Control and Prevention (CDC) reported that more than 75,000 people were enrolled in the WTCHP, including 65,672 responders and 10,067 survivors. To date 35,420 of these individuals have received at least one diagnosis, including cancers, respiratory or digestive illnesses, or some combination thereof, with conditions ranging from mild to severe. Many had presented with cancers at much younger ages than expected, or been diagnosed with multiple cancers. By September 2016, at least 1,000 people had died from illnesses related to their exposure to the toxic cloud that engulfed the area for months after the attack. And experts anticipate that rates of some diseases among people who were exposed to the toxic dust will continue to rise.

Not until 2007 did the New York City Medical Examiner's Office begin to include in the official death count people who died of illnesses related to breathing in World Trade Center dust. The first such victim was a civil rights lawyer, 42-year-old Felicia Dunn-Jones, who died from a chronic lung condition six months after the attacks. By the fifteenth anniversary, the official death toll had reached 2,977.

Some reports have predicted that by the twentieth anniversary, the death toll among those sickened by the dust and debris from Ground Zero will exceed the number of people killed on September 11, 2001. But Dr. Michael Crane, medical director of the WTCHP at Mount Sinai Medical Center, is not one for doomsday prognostication. He said he feels more optimistic now than he has at times past. While he is concerned about rising rates of lymphoma, for example, and he is certainly concerned that some of his patients are gravely ill, Crane is also encouraged by increased options for helping people to stay, and get, well. More recent, "more complete" data suggests to him that "we may have more levers to pull and push" in terms of both treatment options (including new drug regimens effective against multiple myeloma, for example) as well as prevention. Particularly encouraging to him are findings that lifestyle changes can make a big difference. "Diet and exercise seem to be pretty damned good. Exercise has a very positive impact on a lot of cancers. That's very encouraging," he said. "It really isn't all doom and gloom."

Even though he's "suffering greatly for it now," McPhillips still considers his ability to serve at Ground Zero an honor. "I've been blessed. I've been so lucky to be part of this," he said. "We saved a lot of lives that day. We saved a lot of people. I really think that's why I feel like I was lucky. Not that we had a choice, because . . . Well, I guess we did have a choice. We could have just not acted. But that's not in our DNA. . . . If you ask anybody that was down there they'd do it again tomorrow."

———⚬⚬⚬———

In reality however, many responders would wind up never having that chance. In addition to the 414 first responders who lost their lives that day, some 2,000 were injured, some so badly they could no longer serve. Bob Nussberger, the volunteer firefighter who was thrown against a building as the South Tower came down, was among them. September 11, 2001, turned out to be Nussberger's last day of firefighting duty.

At the hospital, doctors found that his back was not, in fact, broken. Just his nose, and six toes, which had been fractured by the tips of his steel-toed boots. He'd suffered a concussion, nerve trauma to his neck, injuries to his shoulders, and significant damage to his ears. Before long, members of the Broad Channel Volunteer Fire Department found him, picked him up by ambulance, and took him home to his wife who'd thought he was dead. The heart attacks he suffered in the months thereafter spelled the end of his career with the fire department. Years later, he would wind up restricted to face-to-face conversations because he counted on lip reading to make out the words.

For others, that day of service was just the beginning. Rich Varela was back at work a few days after that September Tuesday when his company called him to say that the New York City Fire Department had tried to reach him. When Varela called back, he spoke with Tom Sullivan. "We've got your bag down here," Sullivan said and then made a joke. "These are some good checks," he said of the two sizable paychecks they'd found in Varela's bag. "We're gonna cash 'em if you don't come get 'em."

When Varela arrived at the firehouse to pick up his bag, the firefighters from Marine 2 recognized him. "Aren't you the guy that was running around with no shirt helping on the boat?"

"Yep, that was me."

They thanked him, gave him a T-shirt, and before long invited him down for lunch. Over the meal Varela asked Sullivan "'What do you gotta do to become a New York City fireman?"

"Why the hell would you wanna do that after these guys all died?" Sullivan replied.

"I dunno, I just felt like this was the first time in my life I felt like I did something worth a shit," responded Varela. At the time he was miserable at his information technology job. "I couldn't stand it." He had started in the field because his dad was involved in the industry. "I just kinda knew it, and it was good money so I just took it," he explained.

Sullivan gave him a little information, pointed him to a website, and that was it. A while later, Varela called Sullivan with news: "I'm on the list." Sullivan was shocked. "He didn't really think I was gonna go through with it." Today, Varela serves as a firefighter with the FDNY's Engine 28, Ladder 11. "Now, it's funny to have beers with Tom, both in our dress blues," said Varela. "We try to see each other every year, each 9/11, or at least get a phone call in."

When he started with the fire department, Varela had the idea that if he ended up putting in 20 years on the job, he would like to work his last year aboard fireboat *John D. McKean*, to complete the circle. "I didn't know they were going to decommission it," he said. Yet, indeed, 56 years after the boat began its service in 1954,

McKean was retired. In early 2016, it was purchased by a pair of restaurateurs who said they plan to operate it as a museum, of sorts, that will be open for tours led by former firefighters.

Auctioning off the old fireboats to make way for the new was part of the FDNY's post-September 11 efforts to overhaul and expand its maritime operations, revamping its approach to emergencies on the city's waterways and along more than 400 miles of shoreline. Fleet expansion, including both large and small boats, was a significant part of that mission. Fireboats' participation in evacuation and firefighting efforts on September 11 had made clear the essential role of the Marine Division in protecting the region from new perils. "We have the same mission, but the threat environment is very different," explained Assistant Chief Joseph Pfeifer (the former battalion chief who was the first chief to arrive at the World Trade Center after the first plane hit) in his current position as founding director of the FDNY's Center for Terrorism and Disaster Preparedness.

In 2010, two new 140-foot floating fortresses were commissioned, incorporating new technologies, with features including on-board protections against chemical, biological, radiological, and nuclear agents, as well as equipment to support the boats' use as command and control centers. One was dubbed *Fire Fighter II* while the other was named *343*, for the 343 FDNY members who lost their lives on September 11. Their construction (with a price tag of $60 million) paid for by a grant from the Department of Homeland Security and New York City, marked the first major capital investment in new boats for the FDNY marine fleet in more than 50 years. In addition to the powerful 140-footers that can pump 50,000 gallons of water a minute, FDNY added smaller fast and maneuverable rapid-response fire and medical rescue boats designed to improve medical care to victims in surrounding waterways.

Although the boats enlisted on September 11 were part of an aging fleet from a different era, their work providing the only firefighting water available at the trade center site for days following the attacks highlighted the importance of an FDNY division that had long suffered what some considered neglect, and even disrespect. Thus, the influx of new equipment. As one firefighter put it: We're going from *The Flintstones* to *The Jetsons.*

In January 2011, firefighter Tom Sullivan, who retired from Marine 1 later that year, told a *New York Times* reporter that his schedule was a lot fuller than it had been. "Day to day, there is a lot more training, there's new equipment," he said. "There's a lot of upkeep, maintenance, involved. It's a new era in the marine division."

Karen Lacey didn't learn that the *John D. McKean* was the fireboat that had rescued her from the water until more than a decade later. While a number of her Wall Street colleagues quit their jobs on October 1, Lacey was "steadfastly against quitting" out of fear. "I, honest to God, did not blame them," she explained years later. "It was crazy." Still, she was determined to proceed as she and her husband had previously planned: she would quit after receiving her January bonus. As it turned out, downsizing that fall led Lacey's company to offer generous early retirement packages. "The timing of that was just right." And before long she began spending her days raising children on the New Jersey side of the Hudson.

For years, Lacey saved the shirt and skirt she'd worn that day. The gray garments, with bits of paper stuck to them, stiffened, almost as if they had "frozen solid." But time went on. She and her husband moved. When Lacey heard the concerns about asbestos and other toxins in the dust, she tossed the clothes. She didn't want them in the basement where her kids played.

After years of recounting her story at bars and family functions, Lacey decided to volunteer with the 9/11 Tribute Center, a project of the September 11th Families' Association, which "brings together those who want to learn about 9/11 with those who experienced it," offering tours led by guides who share their personal stories of the events of that day. She thought her maritime story would add something to the tour experience.

Not long after the attacks, a coworker who had tried to take shelter from the avalanche of debris commended Lacey for her decision to jump into the river. "I give you a lot of credit," he said. "I would have never done it."

"I give you a lot of credit for standing there," she responded. "I couldn't stand there another minute." People still ask her why she wasn't afraid of the water. "At the time, I was more afraid of being trapped."

Florence Fox, another evacuee who had found herself walking shoeless through Jersey City's streets on that Tuesday morning, struggled to find any sense of normalcy for a long while after the attacks. Though Fox couldn't recall how she had been wearing her hair that morning—maybe cornrows—she did remember that her hair fell out in the fall of 2001. From trauma, she said. Kitten's parents, both psychologists, urged Fox to get counseling, but she didn't. "I am from Africa. We have our own ways," she explained, years later. "And I'm stubborn." Instead, she slept for days.

At some point, she reemerged from her bed. "I had to see Kate, and she needed to see me." She visited the child, but nothing was ever the same. "It was so bad," she explained. "People don't understand. When you go through something like that . . ." she began, her voice trailing off. "I would try and take her to the park and we would try to behave normal but we knew it wasn't." Fox could tell that Kitten felt the same way she did. She could see it in the little girl's face. When they went out in the city, Kitten clung tightly to Fox's hand. The child had always told the nanny who had cared for her since infancy that she loved her, but now she said it with a frequency and an urgency that revealed deep fears. She had trouble sleeping and remained terrified of another attack. "This was a child that was happy, and clever, and full of life," Fox explained. "You could see a lot of that was lost."

An estimated 25,000 children were living or attending school in Lower Manhattan on September 11, 2001. Few studies have yet been published about the mental and physical health impacts of the World Trade Center attacks on children, but investigations into the health of 985 children exposed to Ground Zero dust found that respiratory symptoms present six or seven years after September 11 "were associated with 9/11 dust cloud exposure in younger children and with behavior problems in adolescents."

Kate was among those affected, both by breathing troubles, which manifested as repeat bouts of pneumonia as well as asthma, and by post-traumatic stress disorder. Years of treatment for both conditions have since helped her grow into a thriving teenager. In the years after the attacks, she and her family moved several times within Manhattan. They even tried to resettle back into the Albany Street townhouse. But the memories haunted them, and before long the family relocated to Arkansas.

Tammy Wiggs never expected to make a home in New York City for so long after she'd wound up in the Hudson River. She figured she'd stay for a year or so ("for the proverbial *you fall off of a horse, and you get back on* experience"), but she wound up remaining at Merrill Lynch as an equity trader until moving to Maryland in early 2007. The fifteenth anniversary of the attacks found her living in Baltimore with her husband and three daughters, and working as a senior trader at T. Rowe Price. For all these years, she's mostly kept her story of that morning to herself, only sharing the details with family and very close friends. "It just seems too hard sometimes," she explained.

She still shoulders the survivor's guilt of losing one of her best college friends that day. Wiggs's sister—the one who had evacuated her office in the corner of the Lehman building before World Trade Center tower beams sliced through it—also lives in Baltimore, works in finance, and has three girls. Their families now spend lots of time together.

Each year Wiggs and her sister mark the anniversary together, taking off work to ride horses, take walks, or play golf with their father—steering clear of television and refocusing on family, health, and what truly matters.

CHAPTER 14

September 11, 2016

ON SEPTEMBER 11, 2016, the wooden fencing along the narrow walkway lining the north edge of South Cove looks like it's been replaced a few times over since NYPD Harbor Unit pilot Tony Sirvent instructed his crew to pluck off sections 15 years earlier. Today a man in a red plumbing-and-heating-supply-company T-shirt stands smiling surrounded by five separate fishing rods that rest against the new railings. He chats on a cell phone, surveying his equipment, eyes alert for any movement that might suggest a blue crab has made its way into one of the metal baskets resting on the bottom.

In the shade of the tree-lined pathway on the eastern edge of the cove a woman chases a toddler chasing a ball. The families taking Sunday strolls push fancy strollers—likely later-model versions of the same brands that Sirvent permitted aboard *Harbor Charlie* out of concern for parents and caregivers who otherwise would have arrived on the Jersey side without.

The commonplaceness of weekend activities is interrupted briefly when three men in Coast Guard K-9 Unit ball caps walk three leashed dogs along the perimeter of the cove, a few feet away from a sign that reads, "Dog-Free Park." The working-dog and handler pairs are among the Coast Guard's canine explosive detection teams located across the country. Compared with mechanical methods of explosive detection, canines are considered the most reliable, real-time defense against acts of terrorism.

The teams continue north toward the trade center site. Odds are they began their trek at the Coast Guard Station at the southern tip of Manhattan. If so, they would have just emerged from beneath the canopy of weeping willow trees that encircle the Museum of Jewish Heritage, A Living Memorial to the Holocaust. Before that, they'd have passed the restaurant in Robert F. Wagner Jr. Park that boasts stunning views of the Statue of Liberty where crisp white tablecloths flap in the late-summer breeze. Just south of the restaurant stands the newly renovated Pier A. The nationally registered former municipal pier—designed with a green roof to mirror its across-the-river neighbor Lady Liberty—was recently transformed from derelict to elegantly restored, and opened as a bar and restaurant.

Still farther south, in Battery Park, the battered and torn bronze sphere that stood for three decades as the sculptural centerpiece of the World Trade Center now marks time before its promised return home. Accompanied by an eternal flame ignited on September 11, 2002, in honor of those lost, the sphere served as New York City's interim memorial until the permanent memorial opened in 2011. Now it awaits relocation to a spot in Liberty Park, a new, elevated, one-acre green space overlooking the reflecting pools in the National September 11 Memorial plaza.

Due west of the sphere, another monument stands on a stone breakwater, just offshore, honoring the thousands of U.S. merchant mariners who have died at sea—6,600 of them in World War II alone. The American Merchant Mariners' Memorial, dedicated in 1991, was inspired by an actual historical event. The sculpture depicts three bronze figures atop a sinking vessel. One of them extends an arm into the water in a failed attempt to reach a fourth who struggles beside the boat, that fourth man's outstretched fingers remaining, in perpetuity, just out of reach. Heightening the memorial's haunting realism, the man's head is slowly submerged with each rising tide. Details for the sculpture were drawn from a photograph taken during World War II by Nazi seamen aboard a U-boat who attacked a U.S. merchant ship, then captured the image of their doomed victims.

A short walk farther south through Battery Park's winding pathways leads to the squat brick Coast Guard Recruiting and Services Center at 1 South Street that's now emblazoned with the words "U.S. Department of Homeland Security" in raised letters above the front door. In 2002, the Homeland Security Act divided the Coast Guard's 11 statutory missions, and delineated Ports, Waterways, and Coastal Security (PWCS) as the first homeland security mission. The commandant of the Coast Guard then designated PWCS as the service's primary focus, alongside search and rescue.

Here is the southern tip of Manhattan. The end of the land. Next door to the Recruiting and Services Center stands the new, modern Staten Island Ferry terminal, complete with fast food, coffee, and convenience store options, along with huge display screens flashing announcements and ads. Uniformed security guards, some handling bomb-sniffing dogs, stand watch, scrutinizing everyone who enters. A sign on the wall reports the current Maritime Security Level, a system devised by the Coast Guard "to easily communicate pre-planned scalable responses for credible threats." Today's placard shows a yellow "Level 1," meaning that "minimum appropriate security measures shall be maintained at all times." An adjacent sign announces that this facility is "operated pursuant to the rules and regulations of the United States Coast Guard," and that "failure to consent or submit to screening or inspection will result in the immediate denial of access."

Beneath a real estate ad urging viewers to "#livelivelier," an Automatic Identification System (AIS) display screen shows a chart of the harbor, overlaid with moving icons representing boats and ships. AIS equipment, which sends a unique signature that transmits each vessel's identity, type, position, course, and speed, is now required on essentially all commercial vessels. This information is used variously by vessel watch standers for collision avoidance, by maritime authorities to track and monitor vessel movements, and now by Staten Island Ferry passengers to pass the time as they wait for the next boat.

When the next ferry arrives, disembarking passengers are funneled past a uniformed security force down a corridor toward the street while boarding passengers wait behind glass. At last a ferry worker slides open the double pocket doors, and a flood of people stream out and up the two ramps onto the massive, 6,000-passenger giant *Samuel I. Newhouse*.

Captain James Parese has retired. On September 11, 2016, another captain stands at the east-facing helm, wearing dark shades and a white-collared shirt with black- and gold-striped epaulets. He peers out through the glass with a closed-lipped expression while below him, hundreds, possibly thousands of passengers board the double-ended ferryboat. If you ask the crewman on deck for the captain's name he'll tell you Henry, but nothing more. "It's our policy not to give out that information," he'll explain.

The Staten Island Ferry eases out of Whitehall Terminal's Slip One, and Manhattan recedes into the distance. Clusters of buildings, old and new, reveal the island's ever-changing cityscape. In the foreground, the curved, mirror-sheathed facade of a 1988, wedge-shaped building at 17 State Street reflects white clouds in blue sky. For a brief moment, the green pyramidal crown of the former Bank of Manhattan Building, which in 1930 spent less than a month as the world's tallest, peeks out above 17 State before disappearing once again as the ferryboat pulls farther into the harbor. A blinking white light atop the spire of the nation's new tallest building, 1 World Trade Center, flashes in the blue sky.

Soon the whole mixed-use harbor comes into view. Against a backdrop of blue, white, and orange-and-white striped shipping cranes that hover above stacks of multicolored containers, vessels of all sorts dart about the Upper Bay past others biding their time at anchor. Recreational sailboats and charter yachts pass rows of cargo barges piled high with stone. Water taxis and fast ferries shuttle passengers to and fro just off Liberty Island where, above her muddied gray and green gown, Lady Liberty's polished torch gleams bright in the sun. Statue of Liberty ferryboats make their runs bringing her visitors, and today each ferry is accompanied by two orange-hulled, defender-class Coast Guard response boats, well armed.

It wasn't always this way. Less than a year after the 2001 attacks, the head of the Coast Guard's Strategic Analysis Office, Captain Robert Ross, told *National Defense* magazine that the task of building up Coast Guard security operations was as overwhelming as eating an elephant. "We are trying to carve this elephant up in slices that we can swallow without choking," he said. "We got here through many, many years of neglecting the threats that a lot of people knew were there. . . . We are not going to fix these problems overnight." He went on to acknowledge that the waterways will always present vulnerabilities, and that "we cannot prevent everything."

A decade and a half later, the new captain of the port, Captain Michael Day—who, after a stint as deputy sector commander in San Francisco, assumed command of Sector New York in July 2015—said that the current culture of vigilance combined with an even stronger "unity of purpose and effort" than that which he extolled in 2001 have created a far safer port. Today's security systems are much more integrated across agencies than they were before, Day said, mentioning the daily intelligence briefings that he receives from the NYPD as one example. "There's better shared awareness of what's going on," he explained, adding that the "shared consciousness" that grew out of overcoming the World Trade Center attacks together is rooted in a "common frame of reference," a recognition of the devastation that occurred in this city.

These important, although somewhat intangible, differences between then and now have also been reinforced by the very tangible reality of infrastructure. The Port of New York and New Jersey has received what Day called the "enabling mechanism of fairly robust port security grants." Not only does the Coast Guard have better tools and equipment, it also has better systems in place for addressing security issues with a multiagency approach. And now, for the first time, there is an actual maritime evacuation plan.

A key participant in the development of that plan was the Port Authority's Bethann Rooney who spent years building the port's security infrastructure from the ground up. The port has been completely transformed since the open-docks, "lawn chair" days. Immediately after September 11, the Harbor Ops Committee established a security subcommittee, which Rooney cochaired with the Coast Guard. "Discussions began in earnest about the whole gamut of security and emergency response," Rooney explained.

"We said there were five tenets of security: awareness, prevention, protection, response, and recovery. In the response arena we started looking at, well, what if we had to do this maritime response again and needed to evacuate Manhattan? Informal discussions started to look at after-action reports and critiques of what worked and what didn't work, what could we do better, and how do we formalize that?"

In the midst of those discussions, on a Thursday in August 2003, just after 4 P.M., the largest blackout in U.S. history shut down New York City, along with the whole Northeast. This was not a "calling all boats" situation, Rooney was quick to explain. "The immediacy and urgency to get out of the city wasn't there" in the same way it had been in 2001. Instead, transportation shutdowns caused by the power outages that left commuters stranded in Manhattan stretched the ferry system beyond capacity. The blackout experience led to the formation of a Harbor Ops Maritime Evacuation Subcommittee, which then developed and exercised a plan explicitly for maritime evacuation. The work group collected extensive data about all the various ferry landings and docking options, numbers and capacities of different vessels, as well as their needs and features including carriage limitations and docking requirements, among other considerations. Since then, key port players, including the NYPD, New York City OEM, Coast Guard, Port Authority, public and private ferryboat operators, and others, have participated in the development of, and exercised, the plan that everyone hopes will never be used.

On an ongoing basis, members of the port community continue to periodically gather in a room to address what-if scenarios, responding to a specific situation by marshaling resources represented on paper, white boards, and nautical charts, using magnets, markers, and Post-it notes. "You do it as if it was the real time, the real day," said Rooney.

"So, New York Waterway, it's a Thursday evening, January 26 at 5:53. Massive event in New York City. What do you have to respond? What do you have ready now? Start deploying it. How many assets do you have tied up because you don't have captains? How long will it take you to get those captains and crews there so you can get under way?"

Beyond the maritime evacuation, Rooney explained, port officials exercise all types of scenarios all over the port. Every one of the 180-plus facilities that move people and goods across the dock is required to conduct an exercise each year that tests multiple parts of the facility's specific plan. These drills are mandated by legislation that Rooney worked with Congress to help craft. "We had to go from nothing to complying with a set of international and domestic regulations and trying to figure out, what are the threats that we need to protect ourselves from?"

Still, one key lesson from the September 11 boat lift remains top of mind for people like Captain Day: the success of the evacuation evolved in large part out of its spontaneous, unplanned nature. "What's that old saying?" asked Day. "No plan survives first contact with the enemy."

In keeping with the example set by the former captain of the port Admiral Bennis, Day's leadership approach emphasizes trusting his teams to take in the information available in the moment and use their best judgment about what positive actions will make things better. "I certainly wouldn't want to second-guess what they're doing on scene. That's part of our DNA, part of our [Coast Guard] culture, to push on-scene initiative," he explained. When it comes to planning, as Day sees it, it's crucial that the blueprint for managing any security situation must be focused on presenting *options* rather than *prescriptions*. A critical part of any plan has to include the reality that "we're going to adapt and improvise for the best outcome possible."

That said, systemic improvements to protocols and equipment are still an essential component of increased port security. Since September 11, the Port Authority has invested $38 million in security programs, technology, and infrastructure, as well as instituted all sorts of new policies and procedures. In recognition of the effectiveness of the agency's efforts, in June 2014 the Coast Guard awarded the Port Authority the Rear Admiral Richard E. Bennis Award, which honors an exceptional commitment to the security of the United States and the marine transportation system. "The Port Authority of New York and New Jersey's security program ensures that we can provide a safe and secure environment for our customers and employees while having a robust, dynamic security program at our port facilities," Richard Larrabee said at the time. "Receiving this award is especially meaningful given that we are the first recipients and the award is named after Admiral Richard Bennis, who on 9/11 was the Coast Guard captain of the port here in New York."

The award was established to honor Bennis following his death in August of 2003 from incurable melanoma that had spread to his lungs and his brain. Bennis had already been struggling with cancer when he was called upon to lead the Coast Guard's response in the aftermath of the deadliest terrorist attack on U.S. soil. (In fact, on September 10, 2001, doctors had removed the staples that had sutured the rear admiral's head after brain surgery.) But by March 2002, Bennis's health had seemed to improve, and he had taken a post as chief of the Maritime and Land Security Office within a new agency, the Transportation Security Administration, which would soon become part of the newly established Department of Homeland Security. When he accepted the position to develop strategies to protect the nation's rails, roads, pipelines, and waterways from attacks, he took half a dozen Coast Guard officials with him, including Captain Patrick Harris. Harris served as the Maritime and Land Security Office's first chief of staff until retiring in 2003.

Also lost in the years since 2001 were John Krevey, who offered up his public-access Pier 63 as a dock for dinner boats ferrying passengers, and John Doswell, the maritime event producer who helped him manage the crowds.

Fifteen years after the events that defined so many maritime careers and permanently altered so many lives, Staten Island Ferry riders crowd against the railings to see the New York harbor sights unfold.

They watch as the huge, 310-foot-long ferryboat swings wider than usual, a bit farther out to sea, to make sure to avoid crossing the bow of a Hapag-Lloyd container ship. As the ferry approaches the ship, the massive passenger vessel seems tiny, making it clear just how gargantuan the cargo vessel is. Doubtless, in accordance with maritime law, a Sandy Hook Pilot stationed in the bridge is guiding the ship through the harbor. He or she met the ship before it passed beneath the Verrazano–Narrows Bridge, arriving in a launch that lined up with the vessel, like a fly on an elephant, to position the pilot for a climb up the rope ladder slung over the ship's side.

Designed for secure ocean crossings, the *Houston Express*, from Hamburg, lacks the maneuverability to make sharp turns in a crowded port. So at least two tugs are needed to assist the ship's passage, gently nudging the 1,000-plus-foot-long vessel in a coordinated dance choreographed according to the pilot's commands. Markings labeled "TUG" indicate the spots on the hull where this nudging should occur.

As the ferry takes the ship's stern, the first tugboat comes into view, bringing up the rear. Two newer tugs are already in position on the vessel's port side. They'll provide the thrust necessary to keep the containership on course as it hooks to the west, bound for a New Jersey container port.

On the opposite side of the ferry, before the great sweeping arcs formed by the suspension cables of the double-decked Verrazano–Narrows Bridge, a cluster of barges and a modern articulated tug barge unit (in which the tug is coupled directly to the barge with a hinged connection at the boat's bow) wait at anchor for freed-up dock space or a fair tide.

As the ferry advances toward Staten Island, the expanse of blue-black water to the east is bookended by two clusters of skyscrapers. Manhattan's historic financial district rises on the right while across the Hudson at Paulus Hook, just north of the

Colgate Clock, stands "Wall Street West," so called for its concentration of financial company offices. Jersey City marks the southern boundary of the "Gold Coast of New Jersey," a string of rapidly expanding riverfront towns where a patchwork of developments once struggling with high vacancy rates and canceled projects has sprouted into a grove of high rises. The current land grab that has buyers snatching up property at record-high prices was exactly the real estate boom that Arthur Imperatore Sr. was banking on when he founded New York Waterway in Weehawken.

As the Staten Island Ferry captain lines up the ferry for the final approach into the slip, a crew member's voice, low and thick with a New York accent, rings out over the loudspeaker. He informs passengers, multiple times, that they cannot remain aboard this ferry in hopes of a round trip. "Upon arrival all passengers must leave this boat. This boat will be going out of service and will not be returning back to Manhattan. If you wish to return to Manhattan, exit through the terminal and wait for the next available boat."

The next available boat, *Spirit of America*, is waiting in the adjacent slip. In honor of today's anniversary, the flag mounted on its upper deck flies at half-mast. Doubtless some ferry passengers will take notice. Although they are unlikely to have heard about the role that New York harbor's vessels and crews played in delivering people to safety in the aftermath of the World Trade Center attacks, most tourists and natives alike recognize the significance of this day. Who could forget where they were, whom they were with, and what they were doing on September 11, 2001? It's a date burned into the memories of a generation. But, 15 years later, as is proper, we remember, we memorialize, we pay respects, and then we—including the mariners of New York harbor—get back to work.

Afterword

Fifteen years later, I, too, have learned how to get back to work, making peace with the haunting memories churned up by each anniversary. Today I can more readily perceive the helpers through the horrors. And the helpers... The helpers are what still reverberates a decade and a half later. They have reconstructed my faith in the human soul.

When I returned home to Brooklyn from Ground Zero on September 14, 2001, the earth seemed to have shifted on its axis. My life felt irrevocably altered, but I found it difficult to understand or express exactly what had changed.

The boat on which I served as assistant engineer, retired fireboat *John J. Harvey*, was lauded as a "hero of the harbor," and so, by extension, was the boat's crew. But the classic September 11 hero narrative never sat well with me. It seemed rooted in some arbitrary separation between those who help and those who don't. It seemed to hinge on some diminishment of our human potential.

Many people who showed up to work at Ground Zero—firefighters, ironworkers, engineers, journalists, chiropractors, medics, people from trades of all kinds—did so out of a sense of duty and professional honor. Yet, even those with no obviously applicable expertise possessed skills that could be useful, so they used them. These acts have always struck me as less about heroism and more about pragmatism, resourcefulness, and simple human decency. If you have the wherewithal, you step up.

In her book *A Paradise Built in Hell*, Rebecca Solnit crystallizes perfectly the "resilient, resourceful, generous, empathic, and brave" nature of humans confronting disasters:

> "When all the ordinary divides and patterns are shattered, people step up—not all, but the great preponderance—to become their brothers' keepers. And that purposefulness and connectedness bring joy even amid death, chaos, fear, and loss. Were we to know and believe this, our sense of what is possible at any time might change. . . . Horrible in itself, disaster is sometimes a door back into paradise, the paradise at least in which we are who we hope to be, do the work we desire, and are each our sister's and brother's keeper."

Stepping up, with purpose, even in the absence of a plan, allows us to foster the connectedness that is the very manifestation of humanity.

Acknowledgments

This book is based on the first-person accounts of people who shared with me some of the proudest and most harrowing moments of their lives, some recounting for the very first time their experiences during those fraught hours. I am humbled by their generosity in giving their time and their stories, and I am awed by the resourcefulness, joint action, compassion, and power of New York harbor's maritime community, which was harnessed that day to save countless souls.

Innumerable people helped me locate interview sources, including: Bonnie Aldinger, Brandon Brewer, Andy Brooks, Jim Campanelli, Rose Craig, Dan Croce, John Doswell, Jim Ellis, Leslie Gilbert Elman, Linda Galloway Farrell, Tom Ferrie, Nancy Gamerman, Della Louise Hasselle, Pete Johansen, Patrick Kinnier, Edward Knoblauch, Kaitlin Knoblick, Eric Kosper, Erika Kuciw, Madison Meadows, Harry Milkman, Ellen Neuborne, Rose Newnham, Timothy O'Brien, Kelley O'Connor-Iacobaccio, Kelly Palazzi, Jeanne Park, Jean Preece, Carolina Salguero, Jenna Schnuer, Skipper Shaffer, Sue Shapiro, Rich Siller, Christian Sorensen, Charlie Suisman, Christina Sun, Russell Tippets, Alex Weisler, Lewis Werner, Stephanie Wien, and Fred Woolverton.

I'm indebted to Paul Amico, Pete Capelotti, Huntley Gill, Pat Harris, Chris Havern, Tim Ivory, James Kendra, Bob Lenney, Greg Scharfstein, and Tricia Wachtendorf for answering research questions of all sorts. Instrumental to my reporting were photographs taken/provided by Robert Deutsch, Greg Frietas, Jerry Grandinetti, Greg Hanchrow, Pete Johansen, Angela Krevey, Mike Littlefield, Carolina Salguero, Mort Starobin, Rick Thornton, and Fred Wehner, as well as the officers of the NYPD Aviation Unit.

I'm grateful to Molly Mulhern for dreaming up this project and trusting me to write it, Janet Robbins for patience while I tied up loose ends, and Christopher Brown for shepherding the book to the finish line.

Acknowledgments

Thanks to Kent Barwick, Huntley Gill, Clay Hiles, Mark Kramer, Leslie Meredith, Ellen Neuborne, and Nancy Rawlinson for tremendous grant application help. Crucial financial support came from the Furthermore grants in publishing, a program of the J.M. Kaplan Fund, through the gracious sponsorship of the North River Historic Ship Society. Thanks also to James Gregorio for always being in my corner, and to Anika and The Compound CoWork for positivity and a productive space to write.

I cannot overstate the critical importance of the narrative counsel and cheerleading provided at pivotal moments by Trevor Corson, Anu Partanen, and Ellen Rubin. With their generous feedback and support, both Mark Kramer and Geoff Shandler granted me new ideas and narrative directions at make-or-break moments. Ben Rubin graciously read and commented on portions of this material far more times than anyone should ever have had to. And Nancy Rawlinson rescued the book when it was headed for the rocks. Without her warm-yet-tough coaching and keen editorial eye this book would not exist.

My life was forever changed the day I met my agent Joy Tutela, whose unflagging commitment, wise counsel, and friendship I treasure. There is no fiercer champion. I am so grateful to her and the David Black Agency team.

Thank you to the crew of fireboat *John J. Harvey* (especially my engineering team of John Browne, Jeff Griswold, Jim Travis, Paul Toledano, and Wendy Range) for keeping things running smoothly during my book-related absences. And thank you to all the amazingly loving and diligent babysitters and teachers who gave me the peace of mind that permitted uninterrupted hours of writing.

Finally, I want to thank my family. Mom, Dad, Ellen, and Harold, your support, encouragement, and faith in me have made all things possible. A special thank you to Mom and Molly for heroic hotel-room baby wrangling that granted me the space to study craft. Thank you, Zillin and Jude, for your laughter, which lights the way. And thank you, Ben, for your love, patience, and steadfast support; for covering for me in countless ways; and for your unwavering belief in this project even when I was plagued with doubt. There are no words sufficient to express my gratitude for the chance to walk shoulder to shoulder with you on this journey.

Vessel Participants in the Evacuation

While not comprehensive, this list, compiled by the late Captain John Doswell, includes many of the vessels that participated in the maritime evacuation of Manhattan on September 11, 2001.

ABC-1, Reynolds Shipyard, tug
Abraham Lincoln, New York Waterway, ferry
Adak, United States Coast Guard, island cutter-class patrol boat
Adriatic Sea, K-Sea Transportation Corp., tug
Alexander Hamilton, New York Waterway, ferry
Amberjack V, Amberjack V, fishing boat
American Legion, New York City Department of Transportation, ferry
Anne, Reid Stowe, schooner
Bainbridge Island, United States Coast Guard, island cutter-class patrol boat
Barbara Miller, Miller's Launch, tug
Barker Boys, Barker Marine Ltd., tug
Baleen, Pegasus Restoration Project, historic whaleboat
Bergen Point, Ken's Marine, tug
Bernadette, Hudson River Park Trust, workboat
Blue Thunder, United States Merchant Marine Academy, fishing boat
Bravest, New York Fast Ferry, ferry
Brendan Turecamo, Moran Towing Corp., tug
Bruce A. McAllister, McAllister Towing and Transportation Co. Inc., tug
Capt. John, John Connell, unknown
Captain Dann, Dann Ocean Towing, Inc., tug
Catherine Turecamo, Moran Towing Corp., tug
Chelsea Screamer, Kennedy Engine Company, Inc., tour boat
Chesapeake, unknown, unknown
Christopher Columbus, New York Waterway, ferry
Circle Line VIII, Circle Line/World Yacht, tour boat
Circle Line XI, Circle Line/World Yacht, tour boat
Circle Line XII, Circle Line/World Yacht, tour boat
Circle Line XV, Circle Line/World Yacht, tour boat
Circle Line XVI, Circle Line/World Yacht, tour boat
Coral Sea, K-Sea Transportation Corp., tug
Diana Moran, Moran Towing Corp., tug
Dottie J, United States Merchant Marine Academy, fishing boat
Driftmaster, United States Army Corps of Engineers, drift collection vessel
Eileen McAllister, McAllister Towing and Transportation Co. Inc., tug
Elizabeth Weeks, Weeks Marine Inc., tug
Emily Miller, Miller's Launch, tug
Empire State, New York Waterway, ferry
Excalibur, VIP Yacht Cruises, dinner/cruise boat
Express Explorer, Express Marine, Inc., tug
Finest, New York Fast Ferry, ferry
Fiorello La Guardia, New York Waterway, ferry
Frank Sinatra, New York Waterway, ferry
Franklin Reinauer, Reinauer Transportation Companies, tug
Garden State, New York Waterway, ferry

Vessel Participants in the Evacuation 223

Gelberman, United States Army Corps of Engineers, drift collection vessel
George Washington, New York Waterway, ferry
Giovanni Da Verrazano, New York Waterway, ferry
Gov. Herbert H. Lehman, New York City Department of Transportation, ferry
Growler, United States Merchant Marine Academy, tug
Gulf Guardian, Skaugen PetroTrans Inc., tug
Hatton, United States Army Corps of Engineers, work vessel
Hawser, United States Coast Guard, small harbor tug
Hayward, United States Army Corps of Engineers, drift collection vessel
Henry Hudson, New York Waterway, ferry
Horizon, VIP Yacht Cruises, dinner/cruise boat
Hurricane I, United States Merchant Marine Academy, utility boat
Hurricane II, United States Merchant Marine Academy, utility boat
JC, unknown, unknown
Jersey City Police Emergency Service Unit boat, Jersey City Police Emergency Service, police boat
John D. McKean, Fire Department City Of New York, fireboat
John F. Kennedy, New York City Department of Transportation, ferry
John J. Harvey, John J. Harvey, Ltd., retired Fire Department City Of New York fireboat
John Jay, New York Waterway, ferry
John Reinauer, Reinauer Transportation Companies, tug
Katherine Walker, United States Coast Guard, cutter
Kathleen Turecamo, Moran Towing Corp., tug
Kathleen Weeks, Weeks Marine Inc., tug
Ken Johnson, Interport Pilots Agency, pilot boat
Kevin C. Kane, Fire Department City Of New York, Fireboat
Kimberley Turecamo, Moran Towing Corp., tug
Kings Pointer, United States Merchant Marine Academy, training vessel
Kristy Ann Reinauer, Reinauer Transportation Companies, tug
Lady, Lady Cruise Lines, dinner/cruise boat
Launch No. 5, USCG Auxiliary, retired police launch
Lexington, VIP Yacht Cruises, dinner/cruise boat
Line, United States Coast Guard, small harbor tug
Little Lady, Liberty State Park Water Taxi, ferry
Margaret Moran, Moran Towing Corp., tug
Marie J. Turecamo, Moran Towing Corp., tug
Mary Alice, DonJon Marine Co. Inc., tug
Mary Gellately, Gellately Petroleum and Towing Corp., tug
Mary L. McAllister, McAllister Towing and Transportation Co. Inc., tug
Maryland, K-Sea Transportation Corp, tug
Maverick, United States Merchant Marine Academy, pilot launch
McAllister Sisters, McAllister Towing and Transportation Co. Inc., tug
Millennium, Fox Navigation, ferry
Miller Girls, Miller's Launch, tug
Miriam Moran, Moran Towing Corp., tug
Miss Circle Line, Circle Line-Statue of Liberty Ferry, Inc., tour boat
Miss Ellis Island, Circle Line-Statue of Liberty Ferry, Inc., tour boat
Morgan Reinauer, Reinauer Transportation Companies, tug
Nancy Moran, Moran Towing Corp., tug
New Jersey, New York Waterway, ferry
Ocean Explorer, unknown, unknown
Odin, K-Sea Transportation Corp., tug
Paul Andrew, DonJon Marine Co. Inc., tug
Penn II, Penn Maritime Inc., tug
Penobscot Bay, United States Coast Guard, bay-class icebreaking tug

Peter Gellately, Gellately Petroleum and Towing Corp., tug
New York, Sandy Hook Pilots Association, pilot boat
Port Service, Leevac Marine, tug
Poseidon, United States Merchant Marine Academy, patrol boat
Potomac, unknown, unknown
Powhatan, DonJon Marine Co. Inc., tug
Queen of Hearts, Promoceans / Affairs Afloat, dinner/cruise boat
Resolute, McAllister Towing and Transportation Co. Inc., tug
Ridley, United States Coast Guard, cutter
Robert Fulton, New York Waterway, ferry
Robert Livingston, New York Waterway, ferry
Romantica, VIP Yacht Cruises, dinner/cruise boat
Royal Princess, VIP Yacht Cruises, dinner/cruise boat
Safety III, United States Merchant Marine Academy, utility boat
Safety IV, United States Merchant Marine Academy, utility boat
Samantha Miller, Miller's Launch, tug
Sandy G, Warren George, Inc., unknown
Sassacus, Fox Navigation, ferry
Sea Service, Leevac Marine, tug
Seastreak Brooklyn, SeaStreak America, Inc., ferry
Seastreak Liberty, SeaStreak America, Inc., ferry
SeaStreak Manhattan, SeaStreak America, Inc., ferry
Seastreak New York, SeaStreak America, Inc., ferry
Smoke II, Fire Department City Of New York, fireboat
Spartan Service, Leevac Marine, tug
Spirit of New Jersey, Spirit Cruises, dinner/cruise boat
Spirit of New York, Spirit Cruises, dinner/cruise boat
Spirit of the Hudson, Spirit Cruises, dinner/cruise boat
Stapleton Service, Leevac Marine, tug
Star of Palm Beach, Promoceans / Affairs Afloat, dinner/cruise boat
Sterling, Lady Liberty Cruises, dinner/cruise boat
Storm, United States Merchant Marine Academy, search and rescue vessel
Sturgeon Bay, United States Coast Guard, bay-class icebreaking tug
Susan Miller, Miller's Launch, tug
Tahoma, United States Coast Guard, cutter
Tatobam, Fox Navigation, ferry.
Taurus, K-Sea Transportation Corp., tug
Tender for tugboat *Bertha*, Darren Vigilant, motorboat
Theodore Roosevelt, New York Waterway, ferry
Turecamo Boys, Moran Towing Corp., tug
Turecamo Girls, Moran Towing Corp., tug
Twin Tube, Reynolds Shipyard Corp., tug
Unk, United States Coast Guard, motor lifeboat
Various New York State Department of Environmental Conservation work boats
Various fishing boats
Various Nassau County police/patrol boats
Various New Jersey state and local police/patrol boats
Various NYPD police/patrol boats
Various other vessels
Various Staten Island Ferries
Various United States Coast Guard utility boats
Various United States Coast Guard rigid hull inflatables
Virginia Weeks, Weeks Marine Inc., tug
Vivian Roehrig, C & R Harbor Towing, tug
West New York, New York Waterway, ferry
Wings of the Morning, United States Merchant Marine Academy, utility boat
Wire, United States Coast Guard, small harbor tug
Yogi Berra, New York Waterway, ferry

Notes

PART ONE: The Situation

Chapter One: "It was a jet. It was a jet. It was a jet!"

4   *2,977 people:* Associated Press, "List of 2,977 Sept. 11 Victims," *Daily Herald*, 2/6/17, http://www.dailyherald.com/article/20110909/news/110909868/.
        The 2,977 total includes 2,753 killed at the World Trade Center (as well as three victims who later died of respiratory disease resulting from dust exposure and whose deaths were reclassified as homicides by the New York City Medical Examiner's Office), 184 killed at the Pentagon, and 40 killed on United Airlines Flight 93 in Pennsylvania. It does not include the 19 hijackers.

7   *400,000 to 500,000 civilians:* It is virtually impossible to ascertain exactly how many people were transported off Manhattan by boat. The earliest estimates released by the Coast Guard topped out at 1 million. Later estimates, such as those published in 2003, ranged between 350,000 and 500,000. (Dan Croce, "Attack on New York: The First Response," *Coast Guard Journal of Safety at Sea, Proceedings of the Marine Safety Council*, April–June 2003, 7.)
        A 2002 U.S. Department of Transportation report cited Port Authority estimates that New York Waterway ferries evacuated 160,000 people and 200,000 to 300,000 people were transported aboard other vessels (U.S. Department of Transportation's John A. Volpe National Transportation Systems Center, "Effects of Catastrophic Events on Transportation System Management and Operations: New York City—September 11," 15, 2/6/17, https://ntl.bts.gov/lib/jpodocs/repts_te/14129.htm).
        Perhaps the most precise figures have been compiled by University of Delaware disaster researchers James Kendra and Tricia Wachtendorf, who combined Manhattan daytime population estimates from a 2012 NYU Wagner School report (Mitchell L. Moss and Carson Qing, Rudin Center for Transportation Policy and Management Wagner School of Public Service, New York University, "The Dynamic Population of Manhattan," March 2012) with an analysis of WTC evacuees completed by researchers Rae Zimmerman and Martin F. Sherman (Rae Zimmerman and Martin F. Sherman, "To Leave an Area After Disaster: How Evacuees from the WTC Buildings Left the WTC Area Following the Attacks," *Risk Analysis*, Vol. 31, No. 5, 2011) to conclude that approximately 415,000 people traveled by boat that morning.

8–9   *quick tour of the numbers:* These port statistics represent 1999 figures provided in a report produced annually by the Port Authority of New York and New Jersey. Port Authority representatives explained that its 2000 report was lost in the World Trade Center collapse and that no 2001 report was produced, given the agency's focus on recovery efforts in the aftermath of the attacks. A spokesperson confirmed that the 1999 figures provide an accurate picture of 2001 activity since the percent change between 1999 and 2001 tallies would have been negligible. Source: Michael Day, "Harbor Safety Committees: A Construct for Comprehensive Harbor Stewardship," (*2001*) 2001 International Oil Spill Conference Volume 2. *International Oil Spill Conference Proceedings: March 2001*, Vol. 2001, No. 2, 35-40, 8/17/16, http://ioscproceedings.org/doi/pdf/10.7901/2169-3358-2001-1-35.

11   *prevent vessel collisions:* "Vessel Traffic Services," United States Coast Guard Navigation Center, 2/6/17, http://www.navcen.uscg gov/?pageName=vtsMain.

NOTES

12	*fitting centerpiece:* Brittany Fowler, "Then and Now: How New York City's World Trade Center has changed in the 14 years since the 9/11 terrorist attack," *Business Insider,* 2/6/17, http://www.businessinsider.com/world-trade-center-pictures-before-during-and-after-911-2015-9/#today-cond-nast-and-others-call-the-freedom-tower-home-but-we-will-never-forget-the-twin-structures-that-stood-there-before-17.
13	*50,000 people:* "Effects of Catastrophic Events," 6.
16	*less than a mile away:* Jim Dwyer and Kevin Flynn, *102 Minutes: The Unforgettable Story of the Fight to Survive Inside the Twin Towers,* New York: Times Books, 2005, 48.
16	*8:48:09:* "Manhattan dispatcher audio tape transcript," *The New York Times,* 2/6/17, http://www.nytimes.com/packages/pdf/nyregion/wtctape1.1.pdf.
17	*Hayden later explained:* Peter Hayden testimony, May 18, 2004, videotaped. From National Commission on Terrorist Attacks, *The 9/11 Commission Report,* New York: W.W. Norton & Company, 2004, 291–2.
17	*each single trade center floor:* Dwyer and Flynn, *102 Minutes,* 50.
17	*strictly a rescue:* Peter Hayden testimony.
17	*to reach the upper floors:* Dwyer and Flynn, *102 Minutes,* 51.
18	*without elevators:* Ibid.
18	*"war footing":* Ibid., 56.
18	*Level 4 marshaled:* The 9/11 Commission Report, 291.
18	*largest rescue operation:* Ibid., 293.
19	*flight controllers had learned:* Ibid., 20.
19	*FAA:* Ibid.
19	*NEADS ordered:* Ibid.
20	*16,400 to 18,800 people:* Ibid., 316.
20	*daytime population:* Office of the New York State Comptroller, "The Transformation of Lower Manhattan's Economy," Report 4-2017, September 2016, 1/26/17, 2, http://www.osc.state.ny.us/osdc/reports/rpt4-2017.pdf.
20	*island of Manhattan:* "Effects of Catastrophic Events," 3.
20	*weekdays in 2001:* Ibid., 4.
21	*within minutes:* Dwyer and Flynn, *102 Minutes,* 174.

Chapter 2: "Shut it down. Shut it down. Shut it down!"

22	*Kenneth Summers's choice:* Kenneth Summers, Smithsonian story #6382, The September 11 Digital Archive, 17 December 2003, 6/7/16, http://911digitalarchive.org/smithsonian/details/6382.
23	*more than 60 stores:* "Effects of Catastrophic Events," 6.
27	*mass evacuations:* Dwyer and Flynn, *102 Minutes,* 129.
34	*grown up around:* Phillip Lopate, *Seaport: New York's Vanished Waterfront,* Washington: Smithsonian Books, 2004, 1.
35	*five container facilities:* Author interview with Bethann Rooney, assistant director for the Port Department, Port Authority of New York and New Jersey, 1/26/17.
38	*radio channels:* The 9/11 Commission Report, 301.
38	*transmissions unintelligible:* Ibid., 283.
39	*chiefs in the lobby:* Ibid., 299.
39	*Tactical 1:* Ibid., 301.
40	*"ride heavy":* Ibid., 297.

Chapter 3: "NEW YORK CITY CLOSED TO ALL TRAFFIC"

42	*Port Authority's retort:* Dwyer and Flynn, *102 Minutes,* 58.
49	*fireproofing material:* Ibid., 58 and 67.
49	*finished about 30:* Ibid., 58.

NOTES 227

50	*this heat:* "Debunking the 9/11 Myths: Special Report—The World Trade Center," *Popular Mechanics,* March 2005, 2/6/17, http://www.popularmechanics.com/military/a6384/debunking-911-myths-world-trade-center/.
50	*about 50 minutes:* Dwyer and Flynn, *102 Minutes,* 207.
50	*at 9:13:* The 9/11 Commission Report, 24.
51	*force of the collapse:* Dwyer and Flynn, *102 Minutes,* 20. Steven Ashley, "When the Twin Towers Fell," *Scientific American,* October 2001, 8/12/13, http://www.scientificamerican.com/article.cfm?id=when-the-twin-towers-fell.

PART TWO: The Evacuation

Chapter 4: "I was gonna swim to Jersey."

54	*professional honor:* James Kendra and Tricia Wachtendorf, *American Dunkirk,* Philadelphia: Temple University Press, 2016, 119.
56	*scientists have since calculated:* Dwyer and Flynn, *102 Minutes,* 212.
56	*500,000 ton mass:* Ashley, "When the Twin Towers Fell," http://www.scientificamerican.com/article.cfm?id= when-the-twin-towers-fell.
56	*enough power:* Dwyer and Flynn, *102 Minutes,* 212.
56–57	*actually breathing in:* Joanna Walters, "9/11 Health Crisis: Death Toll from Illness Nears Number Killed on Day of Attacks," *The Guardian,* September 11, 2016, 1/13/17, https://www.theguardian.com/us-news/2016/sep/11/9-11-illnesses-death-toll.
62	*police officer halted:* Dwyer and Flynn, *102 Minutes,* 172.
62	*Captain Paul Conlon:* FDNY Battalion Chief James McGrath, World Trade Center Task Force Interview with FDNY Captain Paul Conlon, File No 9110487, January 26, 2002, 2/6/17, http://graphics8.nytimes.com/packages/pdf/nyregion/20050812_WTC_GRAPHIC/9110487.PDF.
63	*rubbish bin aflame:* Ibid.
64	*top brass:* Dwyer and Flynn, *102 Minutes,* 204.
65	*"like little peanuts":* Mike Magee, *All Available Boats: The Evacuation of Manhattan Island on September 11, 2001,* New York: Spencer Books, 2002, 40.
70	*races around Manhattan:* Zach Schonbrunaug, "Swim Around Manhattan Is Saved From a Future as Murky as Its Waters," *The New York Times,* August 18, 2016, 2/6/17, http://www.nytimes.com/2016/08/19/sports/swim-around-manhattan-is-saved-from-a-future-as-murky-as-its-waters.html.

Chapter 5: "It was like breathing dirt."

72	*giant scissors:* Dwyer and Flynn, *102 Minutes,* 231.
72	*close to a thousand:* Ibid., 229.
73	*22-inch-wide windows:* Evie T. Joselow, "World Trade Center, New York City," R. Stephen Sennott, Ed. *Encyclopedia of 20th-Century Architecture,* v. 3, New York: Fitzroy Dearborn, 2004, 1452.
74	*$90 million:* The 9/11 Commission Report, 280.
74	*enormous development project:* "History of the Twin Towers," The Port Authority of New York and New Jersey, 2/6/17, http://www.panynj.gov/wtcprogress/history-twin-towers.html.
75	*existing railway tunnels:* "PATH Rail History," The Port Authority of New York and New Jersey, 2/6/17, http://www.panynj.gov/about/history-path.html.
75	*in 1962:* Ibid.
77	*face of the towers:* Dwyer and Flynn, *102 Minutes,* 40.
77	*steel spandrel members:* Ashley, "When the Twin Towers Fell," http://www.scientificamerican.com/article.cfm?id= when-the-twin-towers-fell.
81	*85 percent:* The 9/11 Commission Report, 317.

Chapter 6: "We're in the water!"

86	*11 different companies*: Brian J. Cudahy, *Over and Back: The History of Ferryboats in New York Harbor*, New York: Fordham University Press, 1990, 340.
86	*city-run service*: Ibid., 195.
87	*private ferries began*: Ibid., 317.
90	*five-year training program*: Emily S. Rueb, "The Channel Masters of New York Harbor," *The New York Times*, November 17, 2016, 2/6/17, http://www.nytimes.com/2016/11/20/nyregion/at-sea-with-new-york-harbors-channel-masters.html.
91	*Sandy Hook Pilots*: Kendra and Wachtendorf, *American Dunkirk*, 107.
91–92	*OpSail plans*: Ibid., 106

Chapter 7: "Gray ghosts"

98	*building had twisted*: Dwyer and Flynn, *102 Minutes*, 242.
100	*Kate Silverton*: Name has been changed in accordance with the minor child's mother's wishes.
101	*Susan Silverton*: Name has been changed in accordance with source's wishes.
103	*Military Ocean Terminal*: "Brooklyn Army Terminal History," BKLYN Army Terminal, 2/6/17, https://www.bklynarmyterminal.com/building-information/history/.
108	*New Jersey EMS official*: A.J. Heightman, "Exodus Across the Hudson," *Journal of Emergency Medical Services: Out of the Darkness*, September 2011, v. 2, 9/16/16, http://www.jems.com/content/dam/jems/PDFs/Vol%202_New%20Jersey.pdf, 5.
109	*"all frontal burns"*: Heightman, "Exodus Across the Hudson," 11–12.

Chapter 8: "A sea of boats"

115	*I was sent there*: Kendra and Wachtendorf, *American Dunkirk*, 113.
118	*marshaling stations*: P.J. Capelotti, *Rogue Wave: The U.S. Coast Guard on and after 9/11*, Washington, D.C.: U.S. Coast Guard Historians Office, 2003, 21.
118	*team aboard the pilot boat*: Magee, *All Available Boats*, 58.
127	*Robert Fulton debuted*: George Matteson, *Tugboats of New York*, New York: New York University Press, 2005, 1.
127	*observers described*: Kirkpatrick Sale, *The Fire of His Genius: Robert Fulton and the American Dream*, New York: Simon and Schuster, 2002, 120.
128	*steam ferryboats*: Matteson, *Tugboats of New York*, 21.
128	*vessels doubled in size*: Ibid., 23.
128	*limited call*: Ibid., 25.
128–129	*diesel engine*: Ibid., 224.
129	*critical industries*: Ibid., 1.
129	*shift to containerization*: Ibid., 214.
129	*by the early 2000s*: Ibid., 220.
129	*"a sea of tugboats"*: Magee, *All Available Boats*, 66.
133	*27 tugboats evacuated*: Norman Brouwer, "All Available Boats," *Seaport* magazine, Vol. XXXVII, Number 2 and 3, Spring/Summer 2002, 9.

Chapter 9: "I need a boat."

140	*row of grand buildings*: New York City WPA Writers' Project, *A Maritime History of New York*, second edition, New York: Going Coastal Inc., 2005, 193.

143	*prompting the company:* Richard O. Aichele, "A Shining Light in Our Darkest Hour," *Professional Mariner,* #61, Dec/Jan 2002, http://www.professional-mariner.com/March-2007/A-shining-light-in-our-darkest-hour/.
143	*photographs showed thousands:* Brouwer, "All Available Boats," 10.
147	*involved in a relief effort:* David Tarnow, "All Available Boats: Harbor Voices from 9/11," Interview with John Krevey, commissioned by the South Street Seaport Museum, 1/26/17, https://beta.prx.org/stories/2196.
151	*by nightfall:* "Effects of Catastrophic Events," Appendix, 19.
154	*best of our kind:* From written statement by Pat Harris.
158	*10,000 people:* John Erich, "Across the River from NYC, It Was 'John Wayne Time,'" *EMS World,* September 8, 2011, 2/6/17, http://www.emsworld.com/article/10364660/across-the-river-from-nyc-it-was-john-wayne-time.

PART THREE: The Aftermath

Chapter 10: *"We have to tell us what to do."*

162	*rules were broken:* Kendra and Wachtendorf, *American Dunkirk,* 125.
163	*eye for security:* Ibid., 6.
169	*people left stranded:* "Effects of Catastrophic Events," 1.
171	*more than 50,000:* Brouwer, "All Available Boats," 10.
175	*distraught mother: Saved,* Episode 106, Animal Planet, August 2011 (video).

Chapter 11: *"Sell first, repent later."*

179	*trauma hospital:* "Chelsea Piers Plays Major Role in Days Following September 11th," *On The Piers,* Vol. 5, Issue 1, 1.
181	*frozen zone:* Steve Fishman, "Down By the Frozen Zone," *New York,* October 1, 2001, 2/6/17, http://nymag.com/nymetro/news/sept11/features/5201/.
186	*two opposing missions:* Croce, "Attack on New York," 7.
188	*nearly 60,000:* Lynda Richardson, "On the Busy Ferries, It's Steady as He Goes," *The New York Times,* December 19, 2001.
189	*April 2002:* "Effects of Catastrophic Events," 29.
189	*driving pilings:* Magee, *All Available Boats,* 54.
190	*cutters and small boats:* Ibid.

Chapter 12: *"Thanks for your help!"*

191	*new rules:* Croce, "Attack on New York," 8.
193	*small but efficient operation:* Dan Barry, "Determined Volunteers Camped Out to Pitch In," *The New York Times,* September 23, 2001, 2/7/17, http://www.nytimes.com/2001/09/23/nyregion/a-nation-challenged-volunteers-determined-volunteers-camped-out-to-pitch-in.html.
196	*May 2002:* Michael E. Mazzei, *Pier 25: After the Fall,* 2/14/17, https://www.youtube.com/watch?v=-41APXH9YPQ&feature=youtu.be (video)
197	*small harbor:* Magee, *All Available Boats,* 22.
198	*by day's end:* McKinsey & Company, "Improving NYPD Emergency Preparedness and Response," August 19, 2002, 2/6/17, http://web.archive.org/web/20070615060709/http://www.nyc.gov/html/nypd/pdf/nypdemergency.pdf.

Chapter 13: *"They'd do it again tomorrow."*

199	*plume of toxins:* Centers for Disease Control and Prevention, "First Periodic Review of Scientific and Medical Evidence Related to Cancer for the World Trade Center Health Program," July 2011, 7.

230 NOTES

199 *60,000 to 90,000 first responders:* Maoxin Wu, Ronald E. Gordon, Robin Herbert, Maria Padilla, Jacqueline Moline, David Mendelson, Virginia Litle, William D. Travis, and Joan Gil, "Case Report: Lung Disease in World Trade Center Responders Exposed to Dust and Smoke: Carbon Nanotubes Found in the Lungs of World Trade Center Patients and Dust Samples," *Environmental Health Perspectives,* 2010 Apr; 118(4): 1,1/20/17, https://www.ncbi.nlm.nih.gov/pmc/articles/PMC2854726/.

199 *too ill to work:* Department of Health and Human Services, Michael McPhillips testimony, James Zadroga 9/11 Health and Compensation Act of 2010, Public Meeting, March 3, 2011, 64, 2/6/17, https://www.cdc.gov/niosh/docket/archive/pdfs/NIOSH-226/0226-030311-MeetingTranscript.pdf.

201 *monitoring and treatment:* Centers for Disease Control, "WTC Health Program at a Glance, September 2016," 2/6/17, https://www.cdc.gov/wtc/ataglance.html#overall.

201 *more than 75,000:* Ibid.

201 *at least 1,000 people:* Joanna Walters, "9/11 Health Crisis: Death Toll from Illness Nears Number Killed on Day of Attacks," *The Guardian,* September 11, 2016, 1/13/17, https://www.theguardian.com/us-news/2016/sep/11/9-11-illnesses-death-toll.

201 *chronic lung condition:* Anthony DePalma, "For the First Time New York Links a Death to 9/11 Dust," *The New York Times,* May 24, 2007, 2/6/17, http://www.nytimes.com/2007/05/24/nyregion/24dust.html.

201 *official death toll:* Associated Press, "List of 2,977 Sept. 11 Victims," *Daily Herald,* 2/6/17, http://www.dailyherald.com/article/20110909/news/110909868/.

202 *414 first responders:* Inae Oh and Nick Wing, "16 Sobering Numbers that Remind Us to Honor the Sacrifice of 9/11 Responders," *The Huffington Post,* September 11, 2004, 2/6/17, http://www.huffingtonpost.com/2014/09/11/911-first-responders_n_5797398.html.

204 *expand its maritime operations:* Al Baker, "Fire Dept. Revamps Approach to Emergencies on the City's Waterways, or Nearby," *The New York Times,* January 14, 2011, 2/6/17, http://www.nytimes.com/2011/01/15/nyregion/15fire.html?_r=1&pagewanted=print.

204 *same mission:* Ibid.

204 *one was dubbed:* FDNY Annual Report 2012/2013, 7, 2/6/17, http://www.nyc.gov/html/fdny/pdf/publications/annual_reports/2012_annual_report.pdf.

204 *their construction:* Ibid.

206 *estimated 25,000 children:* "9/11 Health: Children," The City of New York, 2/6/17, http://www.nyc.gov/html/doh/wtc/html/children/children.shtml.

206–207 *health of 985 children:* S.D. Stellman, P.A. Thomas, S. Osahan, et. al., "Respiratory Health of 985 Children Exposed to the World Trade Center Disaster: Report on World Trade Center Health Registry Wave 2 Follow-up, 2007–2008," *Journal of Asthma.* 2013 Feb 18, 2/6/17, http://www.ncbi.nlm.nih.gov/pubmed/23414223

Chapter 14: September 11, 2016

210 *sign on the wall:* "U.S. Coast Guard Maritime Security (MARSEC) Levels," U.S. Coast Guard, 2/7/17, https://www.uscg.mil/safetylevels/whatismarsec.asp.

211 *National Defense magazine:* Roxana Tiron, "Port Security Will Improve, But Gradually," *National Defense* magazine, July 2002, 2/6/17, http://www.nationaldefensemagazine.org/archive/2002/July/Pages/Port_Security4047.aspx?PF=1.

Afterword

219 Rebecca Solnit, *A Paradise Built in Hell: The Extraordinary Communities that Arise in Disaster,* New York: Penguin Books, 2009, 3 and 8.

Index

Academy Bus Company, 146
Aldinger, Bonnie, 22–24, 148–149
American Dunkirk (Kendra and Wachtendorf), 162
American Merchant Mariners' Memorial, 209
Amico, Paul, 83–85, 134–136, 186–188, 189
Automatic Identification System (AIS), 210

Baleen, 155–157
Battery, 44, 51, 58, 60, 76, 80, 95, 114, 131–133, 137, 144, 150, 161, 180, 184, 197
Battery Park, 106, 118, 166, 188–189, 209
Battery Park City, 67, 70, 100, 122, 147, 187, 194
Bennis, Richard E., 11–12, 31, 162–163, 185, 186, 215
bodies, falling/jumping, 29, 37, 43, 62, 67, 98, 101
Boeing 707 ad, 42
Borrone, Lillian, 197
Bouley, David, 193
Boulud, Daniel, 193
bridges
    closing of, vi, 46–47, 52, 116, 149
    effect of on commute, 75, 86–87
    reopening of, 191
Brooklyn Army Terminal, 103, 138
building upgrades, 74
Burns, Donald, 18

Campanelli, Jim, 37, 58, 110, 122, 124, 125–126
canine units, 173, 208
Carpathia, 141
Centers for Disease Control and Prevention (CDC), 201
Chartier, William (Bill), 104–105, 198
Chelsea Piers, 140–141, 146, 156, 178, 179–180
Chelsea Piers Sports and Entertainment Complex, 141
Chelsea Screamer, 97–100, 111, 166–167, 178, 192, 195
children, 100, 132–133, 206–207. *see also* Silverton, Kate (Kitten)
Circle Line, 142–143, 145, 153, 167
Coast Guard, responsibilities of, 82, 120

Coast Guard Recruiting and Services Center, 209–210
Code of Federal Regulations (CFRs), 34, 82, 120
Colgate Clock, 173, 217
Colicchio, Tom, 193
collaboration, drills on, 32
communications, 38–39, 108–109, 117–118, 160
Conlon, Paul, 62
containerization, 34–36, 129, 179
Cove to Cove race, 70
Coyle, John, 170–171
Crane, Dr. Michael, 201–202
currents, danger of, 45, 69–70, 83

Day, Michael, 88–89, 90–92, 114–115, 118–120, 160–162, 164–166, 174, 180, 184, 185, 212, 214
deaths. *see also* bodies, falling/jumping
    due to exposure, 200, 201
    official count of, 201–202
debris, removal of, 196
decontamination, 158
DonJon Marine, 187, 196
Dorhn, Glenn, 129–130, 132–133
Doswell, John, 148, 180, 200, 215
Dowling, Robert, 69
Downtown Boathouse, 134
Dunkirk evacuation, 7
Dunn-Jones, Felicia, 201
dust
    contents of, 56
    deaths and illness due to, 199–202, 207
    toxicity of, 152, 199

elevators, lack of operation of, 17–18
emergency preparedness, 81–82
Esposito, Joseph, 18, 28
Excalibur, 27, 136–139, 151, 175–176

F-15 Eagles, 19, 107
FDNY's Center for Terrorism and Disaster Preparedness, 204
Feal, John, 200
FealGood Foundation, 200
Federal Aviation Administration (FAA), 19

231

ferries
   history of, 86–87
   ridership of, 188–189
fifteenth anniversary commemoration, 3–4
Fire Fighter II, 204
fireboats, description of, 6
fireproofing material, 49
fires, effect of, 49–50
Flay, Bobby, 193
fleet expansion, 204–205
Flicker, Perry (Flick), 193–194
Flight 11
   commemoration of, 4
   FAA and, 19
   hijacking of, 19
   impact of, 10, 15, 16, 22
   notification of impact of, 11
   transmission from, 19
Flight 77, impact of, 47
Flight 175
   commemoration of, 4
   Flight 11 transmission and, 19
   impact of, 31, 37–38
41497, 32, 117, 190
Fox, Florence, 100–103, 105–106, 168–170, 175, 206–207
Frank Sinatra, 14, 79
Franklin Reinauer, 130–131
Freitas, Greg, 98–99, 166, 178–179, 195
Fulton, Robert, 86, 127

Ganci, Peter, 64
Gately, Kevin, 87–88
General Store, The, 195
George Washington, 79, 82, 84, 141
Giants Stadium, 146
Gill, Huntley, 120–122, 178–179
Gillman, Billy, 37, 38, 93
Giuliani, Rudolph, 87, 115
Graceffo, Vince, 181
Grandinetti, Jerry, 26–27, 59–60, 61, 136–139, 150, 151, 174, 175–176
Ground Zero, 6, 153, 174, 180, 192, 196, 200–202, 206, 218

Hammitt, Josh, 28–29, 150
Hanchrow, Greg, 46, 47–48, 139–140, 141–142, 143–144, 174, 176–177, 180–184, 191–193
Harbor Charlie, 103, 208
Harbor Operations Committee, 165, 197, 212–213
Harbor Operations Maritime Evacuation Subcommittee, 213
Harris, Patrick (deputy commander), 11–12, 31, 88–89, 119, 162–164, 165–166, 190, 215

Harris, Patrick (on *Ventura*), 9–11, 28–29, 60–61, 145, 149–154, 194–195
Hayden, Peter, 17
Hayes, Timothy, 28
Hayward, 125–126
Haywood, Bob, Jr., 138
Haywood, Bob, Sr., 139, 151
helicopter rescue, impossibility of, 27–28
Henley, William Ernest, 4–5
Henry Hudson, 43, 44–45, 80
Hepburn, Alice, 154–156
Hepburn, Pamela, 154–155, 156–158
Hercules, 128
Holland Tunnel, 75
Homeland Security Act (2002), 210
Horizon, 147
Houston Express, 216
Hudson
   currents in, 45, 69–70
   tides in, 57
Hudson and Manhattan (H&M) Railroad/H&M Hudson Tubes, 75

Ideal-X, 34
illness due to exposure, 199–202
Imperatore, Arthur, Jr., 188–189
Imperatore, Arthur, Sr., 87, 217
infrastructure, rebuilding of, 189–190
International Naval Review and Operation Sail (OpSail) event, 91–92, 119, 187
Intrepid, 167
Ivory, Tim, 121, 125–126, 178

Janice Anne Reinauer, 130
Jersey, 86
Johansen, Peter, 15, 50–51, 84, 135
John D. McKean, 36–39, 55, 57–58, 63, 65–66, 93–96, 109–113, 122–124, 126, 172–173, 174, 203, 205
John J. Harvey, 5–6, 120–122, 124–127, 178, 180, 219
John Reinauer, 130

K-9 Units, 208
Kendra, James, 54, 162, 163
Kennedy, Sean, 97–100, 111, 166–167, 178–179, 192, 195
Koenig, Fritz, 12
Krevey, Angela, 149, 155
Krevey, John, 146–148, 152, 156, 215
Kunz, Gray, 193

Lacey, Karen, 48–49, 67–69, 92–96, 109, 111, 112–113, 167–168, 205–206
LaPlaca, Gina, 24–26, 77–79, 170–171
Larrabee, Richard, 72–74, 76, 184–186, 215
Launch 9, 104

Index

Lexington, 27, 136
Liberty Island, 211
Liberty Park, 209
Liloia, Donald, 14, 15
Lincoln Harbor, 145, 153–154
Lincoln Harbor Yacht Club, 142, 145, 174
Long Slip, 157
Lusitania, 141

Manhattan
   demographics of, 20
   early ferry service for, 86
   number of people in, 20
   swimming around, 69–70
Manhattan Kayak Company, 149
man-overboard drills, 82
marine assistance, codification of, 54
Marine One, 36, 38
Mariner III, 178, 195
Maritime Security Level, 210
Marriott Hotel, 62, 72, 76
mass casualty incident (MCI) protocols, 108
McGovern, Andrew, 89, 91, 165–166, 182
McPhillips, Michael, 13–15, 79, 81, 82–84, 92, 141–142, 160, 192, 199–200, 202
McSwiggins Pub, 172
Metcalf, Ed, 36–37, 38, 40
Metropolitan Transportation Authority (MTA), 20–21
Meyer, Danny, 193
Miano, Dennis, 139, 144, 149, 151–154
Mis Moondance, 3
Morgan Reinauer, 130, 132–133

National Guardsmen, 178
New York, 91, 92, 114–115, 118, 119, 160, 164, 166, 174, 182
New York Circle Line Sightseeing Yachts, 142–143, 145, 153
New York City Department of Transportation (DOT), 20
New York harbor
   data on, 8–9
   size of, 9
   New York Waterway, 13–15, 87–88, 187–188
9/11 Maritime Medal, 200
9/11 Tribute Center, 205
North River, 127
North Tower
   collapse of, 97–98, 111, 134
   fireball in, 22
   first impact to, 10, 15, 16–17, 22
Northeast Air Defense Sector (NEADS), 19
Nussberger, Bob, 61–63, 71–72, 202–203
NYPD Harbor Unit, 208

NYPD's Emergency Service Unit (ESU), 18

Office of Emergency Management, 88, 174, 187, 195
O'Hara, Gene, 181–184, 192
Oldmixon, John, 90
OpSail event, 91–92, 119, 187
overcrowding
   concerns over, 88–89
   reporting of, 161

Palmer, Charlie, 193
Paradise Built in Hell, A (Solnit), 219–220
Parese, James, 55, 129, 171, 211
Parga, Gulmar, 36–37, 39–40, 55–57, 58, 93, 95–96, 111, 112
PATH (Port Authority Trans-Hudson Corporation) trains, 14, 20, 21, 75, 188
Pentagon, 66, 80
Perez, Carlos, 32–33, 51–52, 117–118, 174, 189, 190, 198
Petersen Boat Yard & Marina, 48, 139, 177
Peterson, Ken, 130–131, 133, 196–197
Pfeifer, Joseph, 16–17, 204
Phillips, Mark, 139, 194
Pier 11, 84
Pier 63 Maritime, 147–149, 152–153, 156, 194
Pintabona, Don, 193
Port Authority of New York and New Jersey
   closing of, 35
   containerization and, 34–35
   oversight from, 20
Port Imperial, 87
Port of New York and New Jersey
   cost of shutdown of, 186
   reopening of, 191
Ports, Waterways, and Coastal Security (PWCS), 210
Post, Kenneth "Bob," 33–34, 36
Powell, Tyrone, 104–105
Powhatan, 196
"Praise the Lord and Pass the Ammunition," 194–195
Preece, Jean, 180

radios, 38–39
Rear Admiral Richard E. Bennis Award, 215
Reetz, Chris, 115–117, 156–158, 172
Reinauer Transportation, 129–130, 144
rescue, compulsion to, 54
Responder Program, 201
Rockefeller, David, 74–75
Rokosz, Greg, 142, 145, 146, 174
Ronan, Daniel, 34, 35, 91, 186

Rooney, Bethann, 185, 212–214
Rosenkrantz, Bruce, 148, 155, 156
Ross, Robert, 211–212
Royal Princess, 27, 136, 149, 151–154, 194
rules, breaking of, 161–162
rumors, 110
Rumsfeld, Donald, 110
Ruthie's, 180
Ryan, Chris, 115–117, 156–158, 172

Samuel I. Newhouse, 55, 210–211
Sandy Hook Pilots, 35, 89–91, 119, 164, 165, 174, 216
Sasso, Frank, 158
Scarnecchia, Daniel, 143
SeaStreak, 87
security, changes to, 185, 212–214
September 11 Memorial, 209
September 11th Families' Association, 205
Silverstin, Larry, 13
Silverton, Kate (Kitten), 100–103, 105–106, 168–170, 175, 206–207
Silverton, Susan, 101, 175
Sirvent, Tony, 103–105, 106–107, 181, 197–198, 208
Slattery, Mickie, 109
Solnit, Rebecca, 219–220
South Tower
  collapse of, 51–52, 55–56, 59, 62, 64, 66, 67, 77, 80, 116, 149
  first impact to, 31, 37–38
"Sphere," 209
Spirit Cruises, 46, 139, 141–142, 143–145, 146, 153, 176–177, 181–184
Spirit of America, 217
Spirit of New Jersey, 140, 143–144, 184
Spirit of New York, 140, 143, 176–177, 181, 184, 191–193
Spirit of the Hudson, 140, 143, 184
St. Joseph's Chapel (St. Joe's Supply), 4–5, 193–194
Staten Island Ferry, 8, 87, 171, 189, 210–211, 216
Steamboat Act (1852), 82
steamboats, 86, 127–128
Stephen Scott Reinauer, 129
Suhr, Danny, 62
Sullivan, Tom, 40, 63–65, 112, 113, 123–124, 203–204, 205
Summers, Kenneth, 22, 37–38, 109
supplies, 174, 179–180, 192–193, 195–196
Survivor Program, 201

Thornton, Rick, 43–46, 80–81
Three Forty Three, 204
Titanic, 54, 140–141
transportation shutdowns, 46–47, 52, 116, 149
triage centers, 107–109, 158
tugs, 127, 128–129, 216. *see also* Reinauer Transportation
tunnels
  closing of, 47, 52, 116, 149
  effect of on commute, 75, 86–87
  H&M Railroad and, 75
  reopening of, 188, 191

Varela, Rich, 30–31, 41–43, 65–66, 95, 112–113, 123, 172–173, 203–204
Vazquez, Janer, 145–146
Ventura, 9, 60–61, 145, 149–151, 194
Vessel Traffic Service (VTS), 11, 33–34, 114
VIP Yacht Cruises, 26, 27

Wachtendorf, Tricia, 54, 162, 163
water rescues, 82–83
Weeks Marine, 187, 196
Whyte, Tom, 122–127
Wiggs, Katherine, 111, 207–208
Wiggs, Tammy, 12–13, 48–49, 67–69, 70, 92–95, 109, 111–112, 167–168, 207–208
Wilson, Jaime, 119
Winter Garden, 28, 68, 97, 200
Woods, Greg, 93–94, 95–96
World Financial Center terminal, 84
World Trade Center. *see also* North Tower; South Tower
  bombing of, 17–18, 44
  building upgrades to, 74
  development of, 74–75
World Trade Center Health Program (WTCHP), 200–201, 202

Yamasaki, Minoru, 12, 73

Zadroga 9/11 Health and Compensation Act, James L., 200, 201
Zodiac, 59–60, 61
Zuccotti Park, 26